Essential
Chemistry
for Cambridge IGCSE®

Workbook

For the updated syllabus

Roger Norris

OXFORD

OXFORD
UNIVERSITY PRESS

Great Clarendon Street, Oxford, OX2 6DP, United Kingdom

Oxford University Press is a department of the University of Oxford. It furthers the University's objective of excellence in research, scholarship, and education by publishing worldwide. Oxford is a registered trade mark of Oxford University Press in the UK and in certain other countries

First published in 2016

British Library Cataloguing in Publication Data
Data available

978-0-19-837468-8

3 5 7 9 10 8 6 4 2

Paper used in the production of this book is a natural, recyclable product made from wood grown in sustainable forests. The manufacturing process conforms to the environmental regulations of the country of origin.

Printed in China by Golden Cup

Acknowledgements

®IGCSE is the registered trademark of Cambridge International Examinations.

The publishers would like to thank the following for permissions to use their photographs:

Cover image: Stan Fellerman/Corbis.

Artwork by Q2A Media Services Pvt. Ltd. and OUP.

Although we have made every effort to trace and contact all copyright holders before publication this has not been possible in all cases. If notified, the publisher will rectify any errors or omissions at the earliest opportunity.

Links to third party websites are provided by Oxford in good faith and for information only. Oxford disclaims any responsibility for the materials contained in any third party website referenced in this work.

Introduction

This workbook is designed to accompany the *Essential Chemistry for IGCSE* student book. It is designed to help you develop the skills you need in order to help you do well in your IGCSE Chemistry examination. The book follows the order of the chapters in *Essential Chemistry for IGCSE*. Each page of questions provides additional questions related to each double page in the student book.

The questions focus on the areas you need to know about for your exam:

- Knowledge (memory work) and understanding (applying your knowledge to answer questions about familiar or unfamiliar situations or substances).
- Handling information from data, tables, and graphs.
- Solving problems (including chemical equations and chemical calculations).
- Experimental skills and investigations.

The first 20 units include a range of question types that you will come across in your chemistry examinations:

- Choosing words to complete sentences: you are usually given a list of words to choose from. This will help you learn and remember key facts.
- Putting statements in the correct order or selecting the correct statement from a list.
- Testing your ability to understand chemical formulae and to construct equations.
- Undertaking chemical calculations involving reacting masses, concentration, empirical and molecular formulae, and percentage yield.
- Some questions ask you to interpret data from diagrams, graphs, and tables. Others ask you to interpret the results of investigations that may be unfamiliar.
- Some pages include questions involving extended answers. These will help you organise your arguments and understand the depth of answer that is needed.

Other important features of this workbook that should help you succeed in chemistry include:

- An introductory Language Lab section in each of the first 20 units, which focuses on scientific words. These are often placed in a particular context. Examples include fill-in-the-gap exercises, word searches, and crosswords.
- A unit focusing on language and the importance of identifying key words in questions. This includes vocabulary practice as well as practice in reading and analysing questions.
- A unit focusing on how to make the most of revision time through active revision and mind mapping.
- A unit on mathematics for chemistry. This includes practice in writing formulae, rearranging expressions, working through calculations, and drawing graphs.
- A unit on practical aspects of chemistry including planning an experiment (the selection of apparatus and materials and working safely), measuring, recording data, and drawing graphs. It also includes analysis of results and evaluation. This is followed by a unit suggesting how these aspects of practical chemistry can be applied to projects.
- A selection of IGCSE-style questions of the type that are set in the theory papers will help you to see connections between different parts of the syllabus.
- Full answers are given to all the questions.
- A glossary to help you understand the meaning of important chemical terms.

We hope that the range of differing exercises in this workbook will help you develop your skills in and understanding of chemistry and help you succeed in this subject.

Contents

Contents

Complete the following sentences about the particles in solids, liquids, and gases using words from the list.

close   everywhere   far   fixed   move   regularly   sliding   vibrate

In solids the particles are .... arranged and close to each other. The particles only

.... They do not .... from place to place. In liquids, the particles are not

arranged in a .... pattern and are .... together. The particles move by

.... over each other. In gases, the particles are .... apart and are able to move

.... rapidly.   [8]

1. Box **A** shows the arrangement of 7 particles in a gas. Complete the boxes **B** and **C** to show the arrangement of 16 particles in a solid and 16 particles in a liquid.

   A (gas)   **B** (solid)   **C** (liquid)   [4]

2. Box **Y** shows particles of gas in a container with a plunger.

   a. Draw a diagram in box **Z** to show what happens when the gas is compressed.   [1]

   Y   plunger   **Z**

   gas particles

   b. Use the kinetic particle theory to explain why the pressure in **Y** is less than the pressure in **Z**.

      ....

      ....

      .... [3]

   c. What happens to the pressure when the temperature decreases at constant volume?

      .... [1]

Language lab

Complete the following sentences about melting and boiling using words from the list.

| attraction | boiling | energy | escape | melting | surface | vibrate | weak |

When a solid is heated, the increase in .... makes the particles .... more.

The forces of .... between the particles are weakened. At the .... point,

these forces are .... enough for the particles to be able to move and slide over each

other. When the liquid is at its .... point, the particles have enough energy to

.... from the .... of the liquid.   [8]

1. a. Complete the diagram by writing the names of the changes of state **A**, **B**, **C**, and **D**.

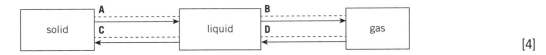

[4]

b. The table shows the melting points and boiling points of three substances.

substance	melting point / °C	boiling point / °C
ethanol	−117	79
methane	−182	−164
naphthalene	81	218

i. Which substance has the lowest melting point? .... [1]

ii. Which substance is a solid at room temperature? Explain your answer?

.... [2]

iii. Which substance is a liquid at room temperature? Explain your answer.

....

.... [2]

c. The diagram shows a heating curve for substance T.

What is the physical state or states of T at the following points?

A .... [1]

B .... [1]

C .... [1]

D .... [1]

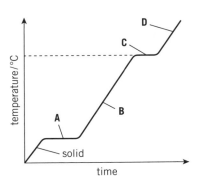

3

Language lab

Search for eight words about the kinetic particle theory in this word square. Words may go upwards, downwards, backwards, or forwards, but not diagonally. Put your answers in the space below.

B	N	E	S	U	F	F	I	D	E
H	J	P	F	A	K	K	U	C	L
B	R	O	W	N	I	A	N	O	C
G	A	S	E	S	N	N	A	L	I
H	N	G	E	N	E	E	N	L	T
I	D	D	D	S	T	T	D	I	R
M	O	V	E	A	I	O	N	D	A
O	M	A	N	P	C	C	Y	E	P

[8]

1. Complete these phrases by writing 'solids', 'liquids', or 'gases' in the gaps.

   a. Particles in .... do not attract each other. [1]

   b. Particles in .... and .... move randomly. [1]

   c. Particles in .... only vibrate. [1]

   d. When particles of unreactive .... collide, they bounce off each other at high speed without any loss of energy. [1]

2. a. Dust particles in still air appear to show Brownian motion.
   Draw a diagram in the space on the right to show a dust particle undergoing Brownian motion.

   [2]

   b. Give one other example of Brownian motion.

   .... [1]

   c. Use ideas of moving particles to explain why dust particles in still air show Brownian motion.

   ....

   ....

   .... [3]

3. The relative molecular masses of three molecules are given.

   ammonia = 17   chlorine = 71   ethane = 30

   Which one of these diffuses most slowly in air? Explain your answer.

   ....

   .... [2]

Language lab

Complete the following sentences about diffusion using words from the list.

changing   gases   hit   liquids   mixed   movement   particles   randomly

In liquids and gases, the .... are constantly moving and .... direction

when they .... other particles. We say that they move .... Diffusion is

the random .... of particles in any direction so that they get .... up.

Diffusion in .... is faster than in .... because the particles move faster

in gases.   [8]

1. A student placed a crystal of a blue dye at the bottom of a beaker of water. After 5 minutes, the crystal disappeared. After 1 day, the solution was blue throughout.

water
dye
crystal
       start   after 5 minutes   after 1 day

    a. State the name of the process occurring when:

       i. the dye changes from a solid to a solution .... [1]

       ii. the colour spreads throughout the water. .... [1]

    b. Use ideas about moving particles to explain the results shown in the diagram.

       ....

       ....

       .... [3]

2. A diffusion experiment is set up as shown.

                            A   B

      cotton wool soaked   glass   white solid   cotton wool soaked
      in ammonia solution   tube   forms here   in hydrochloric acid

Ammonia solution gives off ammonia gas. Hydrochloric acid gives off hydrogen chloride gas. Explain why the white solid forms and why it is closer to **B** than **A**.

....

....

.... [3]

Search for eight pieces of laboratory apparatus in this word square. Words may go upwards, downwards, backwards, or forwards, but not diagonally. Put your answers in the space below.

E	S	R	E	K	A	E	B
L	Y	R	R	A	L	T	A
A	R	O	P	N	O	T	L
M	I	F	I	D	N	E	A
O	N	L	P	E	T	R	N
S	G	A	E	R	R	U	C
T	E	S	T	T	U	B	E
I	N	K	T	E	R	M	H
A	S	R	E	M	I	T	E

[8]

1. The diagram shows four pieces of laboratory glassware.

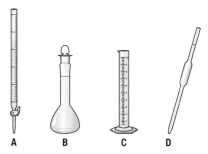

A   B   C   D

a. Name each of these pieces of glassware.

A .... B ....

C .... D .... [4]

b. Which piece of glassware would you use to:

i. Make up a solution of sodium hydroxide accurately? .... [1]

ii. Deliver 25.0 cm³ of hydrochloric acid? .... [1]

2. The diagram on the right shows part of a burette. Where should you position your eye (direction **A**, **B**, **C**, or **D**) to get a precise reading? Ring the correct answer.

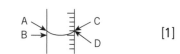

[1]

Complete the following sentences about chromatography using words from the list.

> attraction   filter   locating   mixture   separate   solubilities   spraying

The method of separating a .... of coloured substances using .... paper

is called chromatography. The colours .... if they have different .... in

the solvent and different degrees of .... for the filter paper. Chromatography can

also be used to separate colourless substances. These are shown up after chromatography

by .... the paper with a .... agent.   [7]

1. Paper chromatography can be used to separate a mixture of dyes. Complete the diagram on the right to show the apparatus set-up for chromatography. Label your diagram.   [3]

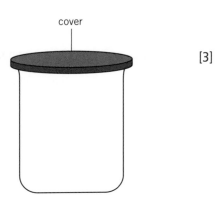

cover

2. A paper chromatogram of some amino acids from a mixture is shown. Two pure amino acids, Ser and Gly, were also run on the same piece of paper.

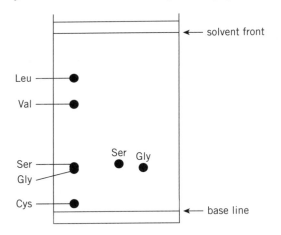

a. Why was the base line drawn in pencil and not in ink?

    .... [1]

b. How many amino acids have been completely separated? .... [1]

c. Which amino acids have not been separated? .... [1]

d. Calculate the R_f value of Val. .... [1]

e. Lysine has an R_f value of 0.14. On the diagram above, draw the approximate position of Lys. Label it Lys.   [1]

Language lab

Complete the following sentences about pure and impure substances using words from the list.

boiling   decreased   exact   impure   increased   pure   range

The melting and .... points of .... substances are sharp. They melt and boil at

.... temperatures. The melting and boiling points of .... substances are not

sharp. They melt over a .... of temperatures. The boiling point of a liquid is

.... if impurities are present. The melting point of a liquid is .... if

impurities are present.   [7]

1. a. Which two of these substances are most likely to be pure?
        Underline the correct answers.

        air   aspirin tablets   orange juice

        oxygen gas   sodium chloride crystals   tap water   [2]

    b. Why should pure substances be used to make a medical drug?

        .... [1]

    c. Seawater is a mixture.

        Suggest a value for the melting point of seawater. .... [1]

2. a. Sulfur melts at 119 °C and boils at 445 °C.

        Draw lines between the boxes on the left and the boxes on the right to complete four sentences.

        ..... melts over a 4 °C temperature range.

        Pure sulfur .... turns to a vapour at 450 °C.

        Impure sulfur .... solidifies at 119 °C.

        ..... has a sharp boiling point.   [2]

    b. Solder is a mixture of tin and lead that is used to join metals.
        The melting point of tin is 232 °C. The melting point of lead is 328 °C.
        Solder melts at 183 °C.

        i. Why does solder have a lower melting point than either tin or lead?

            .... [2]

        ii. Suggest an advantage of the low melting point of solder.

            .... [1]

Link the words **A** to **E** on the left with the descriptions **1** to **5** on the right.

A Decanting	**1** A substance which dissolves in a solvent.
B Filtrate	**2** A mixture formed from a solute and solvent.
C Residue	**3** Pouring off a liquid from a solid which has settled.
D Solute	**4** In filtration, the liquid which goes through the filter paper.
E Solution	**5** In filtration, the solid which remains on the filter paper.

[3]

1. a. i. Complete the diagram by writing the correct labels on the dotted lines.

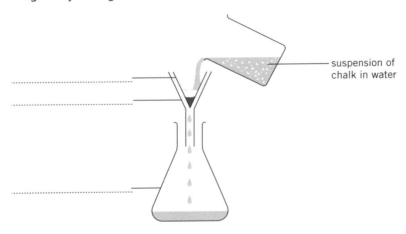

suspension of chalk in water

[3]

ii. On the diagram above label the residue and the filtrate. [2]

2. a. Put these statements about the crystallisation of zinc sulfate in the correct order.

A Filter off the crystals
B Heat the solution to concentrate it
C Dry the crystals with filter paper
D Wash the crystals with a small amount of distilled water
E Leave the solution to cool and form crystals
F By seeing if crystals form on a cold surface
G Check that a saturated solution has formed

Order .... [2]

b. Why should you only use a small amount of distilled water to wash the crystals?

.... [1]

Complete the following sentences about fractional distillation of alcohols using words from the list.

boiling	condenser	further	higher	liquid
lower	receiver	temperatures	vaporised	volatile

There is a range of .... in the distillation column, ... at the

top and .... at the bottom. When .... the more ....

alcohols move .... up the column than the less volatile alcohols. When

the alcohol reaches the .... it changes from vapour to ....

The alcohols are collected one by one in the ...., those with the lower

.... points condensing before those with higher ones. [10]

1. a. i. Label the diagram of the distillation apparatus to show: (i) the distillation flask; (ii) the distillate; (iii) the condenser; and (iv) where cold water enters.

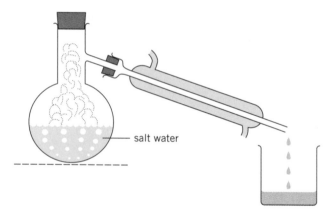

salt water

[4]

   ii. On the diagram above, draw an arrow to show where heat is applied. [1]

   b. i. Explain why this method can be used to separate salt from salty water.

   .... [1]

   ii. Explain why this method cannot be easily used to separate two liquids which have similar boiling points.

   .... [1]

2. Name the separation method you could use to separate:

   a. Sand from a mixture of water and sand. .... [1]

   b. Water from a solution of copper sulfate in water. .... [1]

   c. Two liquids with different boiling points. .... [1]

Language lab

Complete these sentences about atoms using words from the list.

arranged   chemical   electrons   levels   neutrons   nucleus   shells   smallest

Atoms are the .... particles of matter that can take part in a

.... change. Each atom consists of a .... made up of

protons and .... Outside the nucleus are the .... These are

.... in electron .... or energy ....   [8]

1. In 1906, J.J. Thomson suggested a model of the atom shown below.

   How does this model of the atom differ from
   the simple model of the atom we use today?

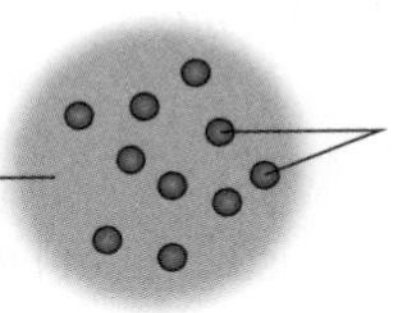

   sphere of positive charge — electrons

   ....
   ....
   .... [3]

2. In the year 1909, scientists fired positively charged particles (alpha particles, He^{2+}) at thin metal foils. The results are shown below.

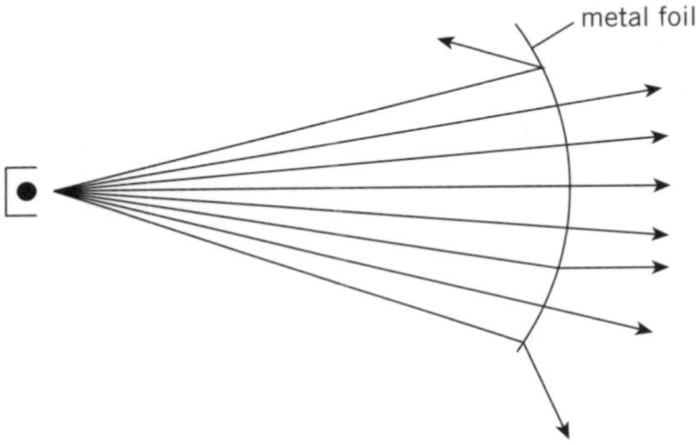

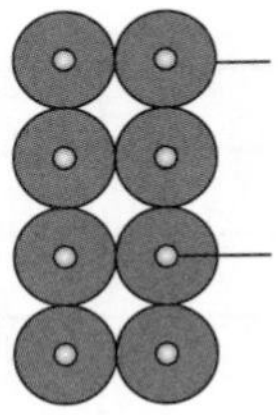

   source of alpha particles — metal foil

   atoms in the metal foil — nucleus — part of the metal foil

   a. How does this experiment show that most of the atom is empty space?

   .... [1]

   b. Explain how this experiment shows that the nucleus has positive charges in it.

   .... [2]

Language lab

Match the words **A** to **D** on the left with the statements **1** to **4** on the right.

A Electrons		1 Particles with no charge
B Neutrons		2 The particles in the nucleus
C Nucleons		3 Particles with a positive charge
D Protons		4 Particles with a negative charge

[2]

1. a. Complete the definition of isotopes using words from the list.

   atoms   compound   electrons   element   mass   molecules   neutrons   protons

   Isotopes are .... of the same .... with the same number of

   .... but different numbers of ....

   [4]

   b. Three isotopes of hydrogen are:

   $_1^1H$ $_1^2H$ $_1^3H$

      i. Determine the proton number of hydrogen. .... [1]

      ii. What is unusual about the isotope $_1^1H$? .... [1]

2. The mass of the isotope phosphorus-32 in a sample and its radioactivity was measured over a number of days. The table shows the results.

time / days	mass of phosphorus-32 / mg	radioactivity / cpm
0	100	200
14	50	100
28	25	50
42	12.5	

   a. Describe how the mass of phosphorus-32 changes with time.

   ....

   .... [2]

   b. Predict the radioactivity at 42 days. .... [1]

3. Give one industrial use of radioactive isotopes.

   .... [1]

Complete these sentences about electrons using words from the list.

arrangement   distribution   electron   Group   one   outer   Period   seven   two

The .... of the electrons in shells is called the electron arrangement or electron

.... An atom of fluorine has nine electrons, ... in the first shell and

.... in the second shell. Atoms of elements in the same .... have the same

number of electrons in their .... shell. As we move across a ...., each atom

has .... more .... in its outer shell than the element before it.   [9]

1. Complete the table to show the electron distribution (electron arrangement) of the atoms shown.

element	number of electrons in an atom	electron arrangement
nitrogen		
oxygen		
fluorine		
neon		
sodium		
argon		
calcium		

[8]

2. Draw the electron distribution (electron arrangement) of these atoms. Show all the electron shells. Draw the electrons in pairs where possible.

aluminium	carbon	chlorine	helium
magnesium	neon	phosphorus	potassium

[8]

Link these phrases to make one sentence describing an element and one sentence describing a compound.

..... types of atom.....	..... one type of atom.....
..... is a substance containing only.....	..... which are chemically.....
..... combined (bonded).	..... which cannot be broken down further.....
..... is a substance containing two or more.....	..... by chemical means.

An element ....

....

.... [1]

A compound ....

....

.... [1]

1. Complete the table to show the difference between a compound and a mixture using words from the list. Some words may be used more than once.

   any   average   combined   definite   different   elements   physical   present   separated

compound	mixture
The .... cannot be .... by .... means.	The substances in it can be .... by .... means.
The properties are .... from those of the .... that went to make it.	The properties are the .... of the substances in it.
The elements are .... in a .... proportion by mass.	The substances can be .... in .... proportion by mass.

[6]

2. The diagram shows six different substances. Each circle represents an atom.
   Classify these as pure elements, pure compounds, or mixtures.

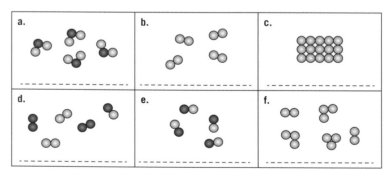

[6]

Language lab

Search for six words that describe the physical properties of iron. Words may go upwards, downwards, backwards, or forwards, but not diagonally. Put your answers in the space below.

[6]

A	L	I	S	E	D	G	R	A	M
L	M	A	L	L	E	A	B	L	E
I	C	E	G	I	N	L	P	U	R
K	I	N	R	T	S	I	I	S	O
D	O	D	A	C	E	E	L	T	L
C	O	N	D	U	C	T	L	R	B
A	S	E	S	D	U	R	E	E	A
X	I	S	U	O	R	O	N	O	S

1. a. The list below gives some properties of metals and non-metals. Underline the properties that are characteristic of **NEARLY ALL** metals.

   brittle   conducts electricity   ductile   dull   high melting point

   high density   insulator   malleable   shiny   strong   [4]

   b. The table gives some properties of diamond (carbon) and sodium.

diamond (carbon)	sodium
melts above 3550 °C	melts at 98 °C
does not conduct electricity	conducts electricity
conducts heat quite well	conducts heat
shatters when hit	malleable

   i. Give one way in which diamond behaves as a typical non-metal.

   .... [1]

   ii. Give one way in which diamond does not behave as a typical non-metal.

   .... [1]

   iii. Give one way in which sodium does not behave as a typical metal.

   .... [1]

2. The table shows some properties of three metals.

metal	density / g/cm³	melting point / °C	electrical conductivity / $\Omega^{-1}m^{-1}$	relative strength
aluminium	2.70	660	0.41	7
copper	8.92	1038	0.54	13
iron	7.86	1535	0.11	21

   Which metal is best for making the body of an aircraft? Explain your answer.

   .... [2]

Language lab

Complete the passage about ionic structures using words from the list.

     alternate   bonds   giant   lattice   negative   positive   regular   strong

A sodium chloride .... is a .... arrangement of .... sodium

ions and .... chloride ions that .... with each other. The ions

are held together by .... ionic .... This structure is called a

.... ionic structure.   [8]

1. Put a ring around the electron arrangements that belong to stable ions or atoms.

    2,8   2,5   2,8,8   2,8,8,2   2   2,8,3   2,8,18,8   [4]

2. Complete the diagrams below to show the electron arrangement of the stable ions.
   Include brackets and charges.

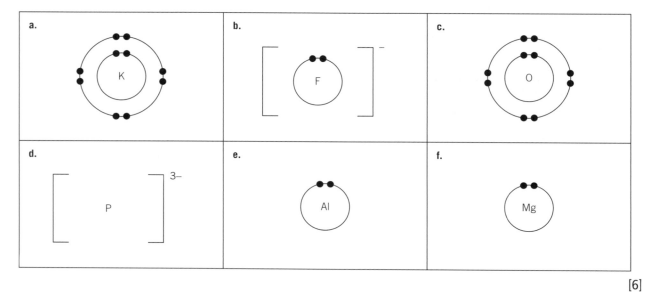

         [6]

3. Complete the ionic structure of magnesium sulfide. Show all the electrons as dots.

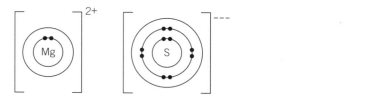

         [3]

Language lab

Link the words **A** to **D** on the left with the correct descriptions **1** to **4** on the right.

A   Molecule

B   Covalent bond

C   Ionic bond

D   Lone pair

1   A bond formed by sharing a pair of electrons

2   A pair of electrons that is not involved in bonding

3   A particle made of two or more atoms held together by covalent bonds

4   The electrostatic attraction between positive and negative ions

[2]

1. Put a ring around the molecules that are diatomic.

CO Cl_2 N_2 N_2O_4 O_2 O_3 P_4 S_8 [2]

2. a. Draw diagrams to show the electron distribution (electron arrangement) in each of these molecules. Show only the outer shell electrons.

i. hydrogen	**ii.** bromine
iii. hydrogen bromide	**iv.** water

[4]

b. i. How many electrons are there around each atom?

.... [1]

ii. What is the significance of this number of electrons for the hydrogen or bromine atoms?

....

.... [2]

Complete the passage about ionic structures using words from the list.

> attraction   electrons   nucleus   strong   two

A covalent bond is formed when .... atoms combine. It forms because of

the .... force of .... between the .... of one atom

and the outer .... of the atom next to it.   [5]

1. Draw dot-and-cross diagrams to show the electron distribution (electron arrangement) of each of these covalent molecules. Show only the outer shell electrons.

oxygen, O_2	nitrogen, N_2
ammonia, NH_3	hydrogen sulfide, H_2S
carbon dioxide, CO_2	ethene, C_2H_4

[6]

Complete the passage about ionic structures and covalent structures using words from the list.

attraction   high   intermolecular   ions   low   strong   weak

Simple covalent compounds have .... melting points because the ....

attractive forces are .... Ionic compounds have .... melting points because of

the .... forces of .... between the positive and negative ....

[7]

1. The table gives some properties of some simple molecular covalent compounds and ionic compounds. Complete the table by writing either 'covalent' or 'ionic' in the last column.

compound	melting point / °C	solubility in water	electrical conductivity when molten	covalent or ionic?
barium oxide	1918	soluble	conducts	
carbon tetrachloride	−23	insoluble	does not conduct	
potassium bromide	734	soluble	conducts	
carbon disulfide	−111	insoluble	does not conduct	
octane	−57	insoluble	does not conduct	

[2]

2. Link the properties **A** to **G** on the left with the correct reasons **1** to **7** on the right.

A Simple molecular compounds have low melting points .....

B Ionic compounds have high melting points .....

C Simple molecular compounds do not conduct electricity .....

D Some simple molecular compounds do not dissolve in water .....

E Ionic compounds conduct electricity when molten .....

F Many ionic compounds dissolve in water .....

G Some molecular compounds dissolve in organic solvents .....

1 ..... because there are no ions or mobile electrons present to conduct.

2 ..... because the molecules cannot form strong enough intermolecular forces with water molecules.

3 ..... because they can form relatively strong bonds with the water molecules.

4 ..... because the forces of attraction between the molecules are low.

5 ..... because they can form relatively strong intermolecular forces with solvent molecules.

6 ..... because the ions are free to move.

7 ..... because there are strong forces of attraction between all the ions.

[3]

Language lab

Search for seven words which refer to diamond or graphite or both. Words may go downwards or forwards, but not backwards, upwards, or diagonally. Put your answers in the space below.

C	O	V	A	L	E	N	T	X	U	A
T	E	T	R	A	H	E	D	R	A	L
P	L	U	T	Y	O	C	R	A	C	L
O	I	D	K	E	Z	P	E	G	A	O
G	A	N	O	R	E	G	I	A	N	T
G	A	D	F	S	Z	S	A	N	E	R
I	P	O	S	S	X	U	M	E	X	O
A	R	R	I	S	B	I	G	I	A	P
D	E	L	O	C	A	L	I	S	E	E

[6]

1. a. Some physical properties, **A** to **F**, are shown below.

   A conducts electricity   **B** does not conduct electricity   **C** hard

   D high melting point   **E** low melting point   **F** soft

   Write the letters of the properties which refer to:

   Diamond .... [1]

   Graphite .... [1]

   Silicon dioxide .... [1]

   b. Which two of diamond, graphite, and silicon dioxide are allotropes? Explain your answer.

   .... [1]

2. Link the observations **A** to **E** on the left with the explanations **1** to **5** on the right.

A Giant covalent structures have a high melting point	**1** ..... because the delocalised electrons are free to move along the layers.
B Graphite conducts electricity	**2** ..... because the weak forces between the layers can easily be overcome.
C Diamond does not conduct electricity	**3** ..... because the carbon atoms are packed closer to each other on average.
D Graphite is soft	**4** ..... because it takes a lot of energy to break the large number of strong bonds.
E Diamond is denser than graphite	**5** ..... because all its electrons are involved in covalent bonding.

[2]

3. Describe the arrangement of the atoms in:

   Diamond ....   Graphite ....   [2]

Language lab

Complete the passage about metallic structure and bonding using words from the list.

across   delocalised   electrons   layers   move   positive   voltage

Atoms of metallic elements are generally arranged in closely packed .... The outer

.... tend to .... away from their atoms to form a 'sea' of ....

electrons around the .... ions. When a .... is applied

.... the metal the delocalised electrons are able to move.   [7]

1. a. Complete the diagram below to show the structure of a metal. Label your diagram.

[4]

   b. Use the information in your diagram to explain why metals are ductile.

   ....

   ....

   .... [3]

2. The bar chart shows the melting point of six successive elements **A** to **F** in the Periodic Table.

   a. One of these elements is a giant covalent structure. Which

      one? .... [1]

   b. Which elements are metals? Give a reason for your
      answer.

      ....

      ....

      .... [2]

   c. Which elements are simple molecules? Give a reason for your answer.

   ....

   .... [2]

The element silicon has the symbol Si. But there are other elements hidden in the word silicon as well if we ignore the lower-case (small) letters or upper-case (large) letters! For example Li is lithium, C is carbon, and O is oxygen.

How many hidden element symbols can you find in the word arsenic (apart from As)? Find six of these and write their symbols and names. For elements with more than one letter, the letters must be together and in the same order as the letters in arsenic.

....

.... [6]

1. **a.** Complete the table to show the valencies (combining powers) of the atoms shown.

						H												
Li											B	C	N	O	F	Ne	
Na	Mg										Al			S	Cl		
K	Ca			transition elements					Zn						Br		

[10]

b. Which atoms in the table lose electrons when they form ions?

.... [1]

c. Which atoms in the table gain electrons when they form ions?

.... [1]

d. Write the formulae of the following compounds by balancing the valencies (combining powers).

i. A compound of H and S ....   **ii.** A compound of B and O ....

iii. A compound of C and S ....   **iv.** A compound of C and Br ....

v. A compound of Ca and N ....   **vi.** A compound of Al and O .... [6]

2. Name compounds **R** to **W**

R MgI_2 .... [1]

S $Sr(OH)_2$ .... [1]

T $FeSO_4$ .... [1]

U $Zn(NO_3)_2$ .... [1]

V $(NH_4)_2CO_3$ .... [1]

W $Ca(HCO_3)_2$ .... [1]

Language lab

Search for the names of ten elements. Words may go downwards or forwards, but not backwards, upwards, or diagonally. Write the names in the space below.

O	N	I	C	K	E	L	U	S
I	I	C	A	L	C	I	U	M
L	E	A	D	L	I	T	R	E
A	I	R	F	O	E	Z	A	N
N	I	B	P	T	Q	N	N	T
B	R	O	M	I	N	E	I	I
T	O	N	L	N	E	O	U	M
E	N	G	W	A	E	N	M	X
R	A	Q	S	U	L	F	U	R

[10]

1. Complete these sentences:

    a. Non-metal atoms form non-metal ions by the .... of electrons.   [1]

    b. Metal atoms form metal ions by ....   [1]

2. **a.** Work out the formulae of compounds **A** to **H** using the list of ions below.

    Al^{3+} Br^- Ca^{2+} Cl^- Fe^{3+} H^+ K^+ Mg^{2+} N^{3-} Na^+ O^{2-} S^{2-}

    A magnesium bromide ....   **B** sodium oxide ....

    C hydrogen chloride ....   **D** aluminium chloride ....

    E potassium nitride ....   **F** calcium sulfide ....

    G aluminium sulfide ....   **H** iron(III) oxide ....   [8]

    b. Work out the formulae of compounds **J** to **Q**. Use the list of compound ions below to help you.

    CO_3^{2-} HCO_3^- NH_4^+ NO_3^- OH^- SO_4^{2-}

    J magnesium nitrate ....   [1]

    K potassium sulfate ....   [1]

    L ammonium nitrate ....   [1]

    M ammonium sulfate ....   [1]

    N calcium hydroxide ....   [1]

    O sodium hydrogencarbonate ....   [1]

    P aluminium nitrate ....   [1]

    Q lithium carbonate ....   [1]

Link the words **A** to **D** on the left with the correct descriptions **1** to **4** on the right.

| A Diatomic | 1 Containing one atom |

| B Monatomic | 2 Substances that are formed in a reaction |

| C Products | 3 Containing two atoms |

| D Reactants | 4 Substances that are used up in a reaction | [2]

1. Write chemical equations for these 'model' reactions.

   a. $O=O$ and

$$H-H \qquad H-O-H$$

   gives

$$H-H \qquad H-O-H$$

   .... [2]

   b. C   and   C   and $O=O$ gives $C\equiv O$ and $C\equiv O$

   .... [2]

2. Balance these equations

   a. $K + Br_2 \rightarrow KBr$

   .... [1]

   b. $Al + O_2 \rightarrow Al_2O_3$

   .... [1]

   c. $Na + O_2 \rightarrow Na_2O$

   .... [1]

   d. $N_2 + H_2 \rightarrow NH_3$

   .... [1]

   e. $Rb + H_2O \rightarrow RbOH + H_2$

   .... [1]

Language lab

Link the words **A** to **D** on the left with the correct descriptions **1** to **4** on the right.

| A Aqueous ion |

| 1 A letter used to tell you if a substance is solid, liquid, gas, or aqueous |

| B Precipitate |

| 2 An ion which does not take part in a reaction |

| C Spectator ion |

| 3 A solid formed when two solutions react |

| D State symbol |

| 4 An ion dissolved in water |

[2]

1. Write down the formulae of the ions present in each of these compounds.

a. NaOH .... and .... [1]   b. $MgCl_2$ .... and .... [1]

c. $Ba(NO_3)_2$ .... and .... [1]   d. $CuSO_4$ .... and .... [1]

e. Al_2O_3 .... and .... [1]   f. $Fe(OH)_2$ .... and .... [1]

2. Write ionic equations for these reactions. In each case, cancel the spectator ions.
The first one has been partly done for you. Where a solid or liquid is formed do not separate its formula into ions.

a. $CuCl_2(aq)$ + $2NaOH(aq)$ $\rightarrow$ $Cu(OH)_2(s)$ + $2NaCl(aq)$

ions   .... + ... $2Na^+ + 2OH^-$ $Cu(OH)_2(s)$ $2Na^+ + 2$ .... [2]

cancel .... + ~~............~~ ~~$2Na^+ + 2OH^-$~~ $Cu(OH)_2(s)$ ~~$2Na^+ + 2$~~ .... [1]

equation $Cu^{2+}(aq)$ + ... (aq) $\rightarrow$ .... (s) [1]

b. $BaCl_2(aq)$ + $MgSO_4(aq)$ $\rightarrow$ $BaSO_4(s)$ + $MgSO_4(aq)$

ions   .... + ... + ... + ... [2]

cancel   .... + ... + ... + ... [1]

equation .... [1]

c. $Cl_2(aq)$ + $2KI(aq)$ $\rightarrow$ $I_2(aq)$ + $2KCl(aq)$

ions   .... + ... + ... [2]

equation .... [1]

Language lab

Complete this sentence about relative atomic mass using words from the list.

A_r atoms   average   carbon-12   twelve

Relative atomic mass (symbol ....) is the .... mass of naturally

occurring .... of an element on a scale on which the .... atom

has a mass of exactly .... units.   [5]

1. a. Complete the table to calculate the relative molecular mass or relative formula mass of the compounds shown.

compound	number of each atom	A_r of atom	M_r calculation
phosphorus trichloride PCl_3	P = Cl =	P = 31 Cl = 35.5	1 × 31 3 × ..... _____ M_r =
magnesium hydroxide $Mg(OH)_2$		Mg = 24 O = 16 H = 1	_____ M_r =
ethanol C_2H_5OH		C = 12 O = 16 H = 1	_____ M_r =
ammonium sulfate $(NH_4)_2SO_4$		N = 14 H = 1 S = 32 O = 16	_____ M_r =
glucose $C_6H_{12}O_6$		C = 12 H = 1 O = 16	_____ M_r =

[10]

b. Calculate the relative formula mass of these compounds.

i. $Al_2(SO_4)_3$ .... [1]

ii. $Co(NO_3)_2$ .... [1]

iii. $Cr(CO)_6$ .... [1]

Language lab

Link the words **A** to **D** on the left with the correct descriptions **1** to **4** on the right.

A Avogadro constant	**1** The ratio of each reactant and product in the equation
B Molar mass	**2** The number of atoms, ions, or molecules in a mole of atoms, ions, or molecules
C Mole	**3** The mass of a mole of substance in grams
D Stoichiometry	**4** The amount of substance that that has the same number of atoms as 12 g of ^{12}C

[2]

1. Use simple proportion to do these calculations about reacting masses.

   When 48 g of magnesium are burnt completely in oxygen, 80 g of magnesium oxide are formed.

   $$2Mg \quad + \quad O_2 \quad \rightarrow \quad 2MgO$$

   a. Complete the calculation to show the mass of magnesium oxide formed when 12 g of Mg are burnt.

      .... × 80 = ... g   [2]

      ....

   b. What mass of magnesium is needed to form 8 g of magnesium oxide?

      .... [1]

   c. What mass of magnesium oxide is formed when 168 g of magnesium are burnt?

      .... [1]

2. Complete the table using the A_r values below.

   C = 12, Ca = 40, H = 1, O = 16, P = 31, S = 32

element or compound	formula mass, M_r	mass taken / g	number of moles
O_2	32	4	
NaCl		11.7	
$CaSO_4$		27.2	
P_2O_5			0.4
CO_2			0.1
P_4		86.8	
CH_4			24.0

[13]

Complete these sentences about mole calculations using words from the list.

adding   atomic   atoms   dividing   moles   number   relative

To find the .... of moles of a compound, we need to know the mass of compound

taken and the .... molecular mass of the compound. The relative molecular mass is

found by .... together the relative .... masses of all the .... (or ions)

in the compound. The number of .... is found by .... the mass of

compound taken by the relative molecular mass.   [7]

1. Complete this equation to show the reacting masses.

$$4PH_3 \quad \rightarrow \quad P_4 \quad + \quad 6H_2$$

$4 \times$ .... $\rightarrow$ $4 \times$ .... $+$ $6 \times$ ....   [1]

.... g $\rightarrow$ .... g $+$ .... g   [1]

2. Use the number of moles shown in the equation below to answer the questions which follow.

$$I_2O_5 \quad + \quad 5CO \quad \rightarrow \quad I_2 \quad + \quad 5CO_2$$

M$_r$ *values*   334   28   254   44

a. How many moles of CO react with 1 mole of I_2O_5? .... [1]

b. How many moles of CO_2 are formed from 1 mole of CO? .... [1]

c. How many moles of I_2 are formed using 10 moles of CO? .... [1]

d. What mass of CO_2 is formed using 20.04 g of I_2O_5 and excess CO?

.... [3]

e. What mass of I_2 is formed using 21 g of CO and excess I_2O_5?

.... [3]

3. 168 g of iron reacts with excess oxygen to form 232 g of an oxide of iron.

A$_r$ values: Fe = 56, O = 16

a. Calculate the mass of oxygen in the iron oxide. .... [1]

b. Calculate the moles of oxygen in the oxide. .... [1]

c. Calculate the moles of iron in the iron oxide. .... [1]

d. Deduce the formula of this oxide of iron. .... [2]

Complete these sentences about gas volumes using words from the list.

conditions   dm³   mole   molecules   number   oxygen   temperature

At room .... and pressure, one .... of any gas occupies 24 ....

(24 000 cm³). Because there is always the same .... of moles, there is also the same

number of .... in a given volume. So, 50 cm³ of chlorine and 50 cm³ of

.... under the same .... contain the same number of molecules.   [7]

1. a. Complete the calculation to show the percentage by mass of carbon in ethane, C_2H_6.

$$\frac{2 \times \text{............}}{(\text{...} \times 12) + (\text{...} \times 1)} \times 100 = \text{.................} \%$$ [2]

   b. Calculate the percentage by mass of nitrogen in ammonia, NH_3. Show your working.

.... %   [2]

   c. Calculate the percentage by mass of calcium in calcium carbonate, $CaCO_3$. Show your working.

.... %   [2]

2. Complete the table to show the mass, moles, or volume of different gases.

gas	M_r of gas	mass of gas / g	moles of gas / mol	volume of gas / dm³
ammonia	17	8.5		
oxygen	32			480
carbon dioxide	44	3.08		
hydrogen chloride		292	8	
ethane	30			3

[10]

Language lab

Complete this relationship using words from the list.

impure   mass   percentage   pure

$$\text{.................. purity} = \frac{\text{.............. of .... product}}{\text{mass of .... product}} \times 100$$ [4]

1. 18 g of limestone reacted with excess dilute hydrochloric acid. 3840 cm³ of carbon dioxide were formed at r.t.p. Work through the calculation to find the percentage purity of this sample of limestone, which is impure calcium carbonate, $CaCO_3$. Give your answer to 2 significant figures.
   A_r values: C = 12, Ca = 40, O = 16

$$CaCO_3(s) \quad + \quad 2HCl(aq) \quad \longrightarrow \quad CaCl_2(aq) \quad + \quad CO_2(g) \quad + \quad H_2O(l)$$

   a. Molar mass of calcium carbonate = ... g/mol   [1]

   b. Volume of CO_2 in dm³ = ... dm³   [1]

   c. Moles of CO_2 = ... = ... mol   [1]

   ....

   d. Moles of $CaCO_3$ in limestone sample = ... mol   [1]

   e. Mass of $CaCO_3$ in limestone sample = ... g   [1]

   f. Percentage purity = ... %   [1]

2. Methyl benzoate can be prepared by reacting methanol with benzoic acid.

$$CH_3OH \quad + \quad C_6H_5CO_2H \quad \longrightarrow \quad C_6H_5CO_2CH_3 \quad + \quad H_2O$$
$$\text{methanol} \qquad \text{benzoic acid} \qquad \text{methyl benzoate}$$

   When 24.4 g of benzoic acid is reacted with excess methanol, 25.84 g of methyl benzoate is produced. Calculate the percentage yield of methyl benzoate. A_r values: C = 12, H = 1, O = 16

   a. Molar mass of benzoic acid = ... g/mol   [1]

   b. Moles of benzoic acid = ... mol   [1]

   c. Moles of methyl benzoate expected (if 100% yield) = ... mol   [1]

   d. Molar mass of methyl benzoate = ... g/mol   [1]

   e. Mass of methyl benzoate expected (if 100% yield) = ... g   [1]

   f. Percentage yield = ... %   [1]

Language lab

Complete these sentences about empirical and molecular formulae using words from the list.

atoms   combine   compound   each   molecule   simplest

The molecular formula of a .... shows the number of .... type of atom in one

.... The empirical formula shows the .... ratio of .... that

....   [6]

1. Complete the following calculations to find the empirical formulae.

   A compound of lead and chlorine contains 20.7 g of lead and 14.2 g of chlorine.
   A_r values: Pb = 207, Cl = 35.5

   moles of Pb = ... mol   moles of Cl = ... mol   [1]

   Divide by   Pb   Cl   ....
   lowest number
   of moles   ....   [1]

   Result of division = ...   = ...

   Simplest ratio .... So empirical formula is ....   [2]

2. Write the empirical formula of the following compounds whose molecular formula has been given.

   a. Hydrogen peroxide, H_2O_2 Empirical formula .... [1]

   b. Antimony(III) oxide, Sb_4O_6 Empirical formula .... [1]

   c. Butane, C_4H_{10} Empirical formula .... [1]

3. Complete the table to deduce the empirical formula mass and molecular formulae of compounds **A** to **C**.
   A_r values: C = 12, Cl = 35.5, H = 1, O = 16, P = 31, S = 32

empirical formula	empirical formula mass	relative molecular mass, M_r	molecular formula
A P_2O_3		220	
B SCl		135	
C CH_2O		60	

                                                                                                                    [6]

Language lab

Complete this relationship using words from the list.

amount   dm³   moles   solute   volume

concentration in .... per dm³ = ... of .... in moles   [5]

.... in ....

1. Complete the table to show the moles and mass of solute and the concentration.

solute	M_r of solute	mass of solute / g	volume of solution cm³ or dm³	concentration of solution / mol/dm³
sodium hydroxide	40	8	250 cm³	
silver nitrate	170		200 cm³	0.5
copper(II) sulfate	160	40		0.125

[3]

2. Work through this calculation to find the concentration of a solution of sodium hydroxide when 25.0 cm³ of a solution of sodium hydroxide is exactly neutralised by 12.2 cm³ of sulfuric acid of concentration 0.100 mol/dm³.

a. moles of acid = ... × $\dfrac{............................}{1000}$ = ... mol H_2SO_4 [1]

b. The equation for the reaction is: $2NaOH + H_2SO_4 \rightarrow Na_2SO_4 + 2H_2O$

    i.   How many moles of NaOH react with 1 mole of H_2SO_4? ....   [1]

    ii. How many moles of NaOH are needed to react with the amount (in mol) of H_2SO_4 you calculated in part **a**?

    .... [1]

c. Calculate the concentration of NaOH in the sodium hydroxide solution.

[2]

Use the clues below to do this crossword.

Across. **1.** Break down

5. Conducts current in or out of the electrolyte

6. A charged atom or group of atoms

7. A power source

Down. **2.** Negative electrode

3. Positive electrode in reverse

4. Electrodes + electrolyte make an electrochemical .... [7]

1. Complete the diagram of an electrolysis cell by labelling the anode, the cathode, the electrolyte, and the power supply.

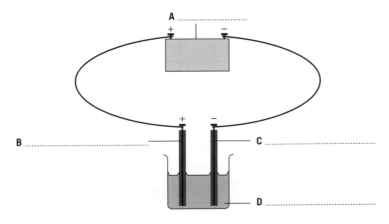

A ....

B ....   C ....

D .... [4]

2. What is meant by these terms?

a. Electrolysis ....

.... [2]

b. Electrolyte .... [1]

3. Complete the table to show the electrode products and observations at the anode when various substances are electrolysed using graphite electrodes.

electrolyte	cathode (−) product	anode (+) product	observations at the anode
molten zinc bromide			
molten magnesium chloride			
molten calcium oxide			
molten lead iodide			

[12]

Language lab

Join up these fragments to form a sentence describing why hydrogen is sometimes formed when aqueous solutions are electrolysed.

.... hydrogen ions in water, ... is more reactive than hydrogen, ... If a metal ....

.... hydrogen, arising from .... bubbles off. .... the metal ions stay in solution and ....

....

.... [2]

1. Complete the table to show the electrode products and observations at the anode when various substances are electrolysed using graphite electrodes.

electrolyte	cathode (−) product	anode (+) product	observations at the anode
concentrated KCl(aq)			
dilute H_2SO_4(aq)			
very dilute NaCl(aq)			
concentrated HCl(aq)			

[12]

2. The diagram shows a cell used for electrolysing brine.

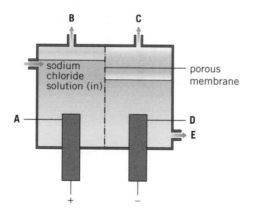

a. Give the names of the gases collected at

B .... C .... [2]

b. Which letter in the diagram represents the anode? .... [1]

c. Explain why oxygen is not given off at the anode.

....

.... [2]

Language lab

Complete these sentences about electrolysis using words from the list.

anode   cathode   electrons   lose   move   oxidation   reduction

During electrolysis, positive ions move towards the .... where they gain ....

This is a .... reaction. Negative ions .... to the .... where they

.... electrons. This is an .... reaction.   [7]

1. Complete these diagrams to show:

   • The movement of ions during electrolysis by drawing arrows.

   • What happens to the ions in terms of electron loss or gain to or from the electrodes.
     (Show this by curly arrows.)

   a.   molten zinc bromide

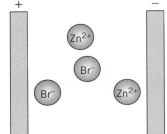

   b.   dilute sulfuric acid

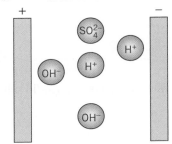

   c.   concentrated hydrochloric acid

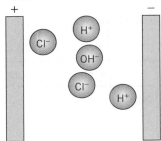

   d.   aqueous copper(II) sulfate

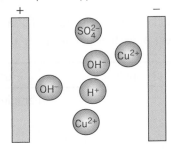

                                                                                        [8]

2. Complete these half-equations for the reaction at the electrodes.

   a. $Zn^{2+} + \ldots \rightarrow Zn$ [1]

   b. ...... $Cl^- \rightarrow$ .... + ...   [2]

   c. ...... $H^+ +$ .... $\rightarrow$ ....   [2]

   d. $Al^{3+} +$ .... $\rightarrow$ ....   [1]

Language lab

Complete these sentences about the purification of copper using words from the list.

anode   atoms   cathode   copper   electrolyte   gain   impure   ions   solution

The electrolysis cell has an .... strip of copper as the .... and a pure strip of

.... as the cathode. The .... is a solution of copper(II) sulfate. At the anode,

copper .... lose electrons and go into .... as copper(II) .... At the

cathode, copper(II) ions .... electrons and are deposited on the .... as

copper atoms. [9]

1. **a.** Complete these equations for the reactions at the electrodes when copper is refined using copper electrodes dipping into aqueous copper(II) sulfate.

   Anode: $Cu \rightarrow$ .... + ... e^- [2]

   Cathode: .... + ... $\rightarrow Cu$ [2]

   b. Explain why the aqueous copper(II) sulfate does not get lighter in colour during this electrolysis.

   ....

   .... [2]

2. Copper(II) sulfate can be electrolysed using graphite electrodes or copper electrodes. Complete the table to show what happens at each electrode and to the electrolyte.

what happens….	using graphite electrodes	using copper electrodes
to the mass of the electrodes?	anode: cathode:	anode: cathode:
to the appearance of the electrodes during electrolysis?	anode: cathode:	anode: cathode:
to the electrolyte? (Give any observations.)		

[10]

Complete these sentences about electroplating using words from the list.

cathode   electrolyte   electroplated   ions   metal   negative   plating   pole   power

The object to be .... is connected to the .... pole of the ....

supply. The object becomes the .... A strip of the plating .... is connected

to the positive .... of the power supply. The .... metal is the anode. The

.... is a solution containing .... of the plating metal.   [9]

1. A nickel jug can be electroplated with silver. Look at the diagram on the right and then answer the questions.

source of electricity

a. On the diagram label the anode, **A**, the cathode, **C**, and the electrolyte, **E**.   [2]

b. What happens to the cathode as the electroplating proceeds?

    ....

    ....

    ....

    .... [2]

c. Complete these equations for the reactions at the anode and cathode.

    Anode: Ag $\rightarrow$ .... + ...   [1]

    Cathode: .... + ... $\rightarrow$ ....   [1]

2. a. Explain why plating an iron object with tin prevents the iron from rusting.

    ....

    .... [2]

    b. Give one other reason why objects are electroplated.

    .... [1]

Search for the names of six substances related to the manufacture of aluminium by electrolysis of aluminium oxide. Words may go downwards or forwards, but not backwards, upwards, or diagonally. Write the names in the space below.

O	C	R	Y	O	L	I	T	E
X	A	L	U	M	I	N	A	R
Y	Z	B	X	X	F	D	A	P
G	R	A	P	H	I	T	E	I
E	P	U	N	O	M	B	R	N
N	A	X	O	C	A	A	O	G
N	F	I	C	A	R	B	O	N
E	R	T	P	L	O	V	D	I
D	Y	E	A	R	A	N	C	I

[6]

1. The diagram shows an electrolysis cell used to extract aluminium from aluminium oxide.

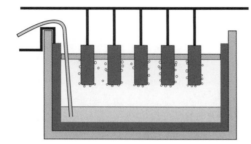

On the diagram above, label the electrolyte as **E**, the cathode as **C**, the anode as **A**, and the molten aluminium as **M**. [4]

2. Explain why the electrolysis mixture is mainly cryolite and calcium fluoride with about 5% aluminium oxide.

....

....

.... [3]

3. **a.** Complete these equations for the reactions at the cathode and anode.

    i. Al^{3+} + ... $\rightarrow$ .... [2]

    ii. .... O^{2-} $\rightarrow$ .... + ... [2]

    b. Construct the overall equation for this electrolysis.

    .... [2]

Join up these fragments to form a sentence about ceramics.

..... because they resist the flow .....   Ceramics are useful .... they have very high .....

..... insulators, not only .... melting points.   ..... of electricity, but also because .....

....

.... [2]

1. a. i. Label the diagram to show the apparatus used to show whether or not a solid
           conducts electricity.   [2]

      ii. On the diagram above put an arrow to show the direction of flow of the electrons.   [1]

2. Aluminium with a steel core is used in high-voltage power cables.

   a. Give two properties of aluminium that are related to this use.

      .... [2]

   b. Give two properties of steel that are related to this use.

      .... [2]

3. Link the phrases **A** to **D** on the left with the phrases **1** to **4** on the right.

| A   Molten sodium chloride conducts electricity ..... | 1   ..... because there are free electrons which move when a voltage is applied. |

| B   Metals conduct electricity ..... | 2   ..... because the ions are not free to move. |

| C   Sulfur does not conduct electricity ..... | 3   ..... because the ions are free to move. |

| D   Solid sodium chloride does not conduct electricity ..... | 4   ..... because none of the electrons is free to move. |

                                                                                                      [2]

Link the words **A** to **D** on the left with the best descriptions **1** to **4** on the right.

A Chemical change		**1** Heat is given out
B Endothermic		**2** No new substances are formed
C Exothermic		**3** Heat is absorbed
D Physical change		**4** New substances are formed

[2]

1. Underline the physical changes.

   Burning magnesium in air   Separating iron from sulfur using a magnet

   Rusting of iron   Melting zinc

   Distilling plant oils from a mixture of plant oils and water   [3]

2. Describe these changes as either exothermic or endothermic.

   a. The decomposition of copper(II) carbonate by heating. .... [1]

   b. Burning paraffin. .... [1]

   c. Your tongue gets cold when you put sherbet on it. .... [1]

   d. The temperature of the solution rises when concentrated hydrochloric acid is diluted.

   .... [1]

3. **a.** Complete these energy-level diagrams for an exothermic and an endothermic reaction. Include an arrow in each diagram in the correct position.

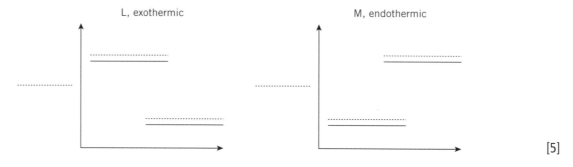

   [5]

   b. Explain why energy-level diagram **L** represents an exothermic reaction.

   ....

   .... [2]

Language lab

Complete these sentences about electrolysis using words from the list.

     absorbed   **bonds**   **greater**   **new**   **out**   **reactants**   **released**

In an endothermic reaction, the energy .... in bond breaking is .... than the

energy .... when new .... are formed. In an exothermic reaction, the energy

given .... when .... bonds are formed is greater than the energy taken in

when the bonds in the .... are broken.   [7]

1. Complete the table to calculate the energy change when methane reacts with excess oxygen to form carbon dioxide and water.

$$\text{H}-\overset{\overset{\textstyle H}{|}}{\underset{\underset{\textstyle H}{|}}{\text{C}}}-\text{H} \ + \ 2 \ \text{O=O} \ \rightarrow \ \text{O=C=O} \ + \ 2 \ \text{H-O-H}$$

Bond energies in kJ/mol: C–H 413, O=O 498, C=O 805, O–H 464

bonds broken (endothermic +) / kJ/mol	bonds formed (exothermic −) / kJ/mol
4 × (C–H) = 4 × 413 = ...	
.... × (O=O) = ... = ...	
Total	

Overall energy change = (+ ....) + (− ....) = ... kJ/mol   [5]

2. Methane burns in excess air to form carbon dioxide and water.

$CH_4(g) \ + \ 2O_2(g) \ \rightarrow \ CO_2(g) \ + \ 2H_2O(l)$ Energy change = −818 kJ / mol propane

Draw a labelled energy-level diagram for this reaction.

     [3]

Language lab

Link the energy sources **A** to **D** on the left with the descriptions **1** to **4** on the right.

A Coal	**1** This is largely methane, the simplest hydrocarbon
B Natural gas	**2** A black liquid which is distilled to give a variety of fuels such as gasoline
C Petroleum	**3** A radioactive element extracted from the ore pitchblende
D Uranium	**4** A black solid formed from the decay of plants in the absence of oxygen

[2]

1. The table shows the temperature changes produced by burning the fuels **A**, **B**, **C**, and **D** underneath a copper can filled with 100 cm^3 of water.

fuel	amount burnt / g	initial temperature of water / °C	final temperature of water / °C	density / g/cm^3
A	2.0	20	25	0.90
B	1.0	18	21	0.55
C	4.0	21	29	0.87
D	3.5	19	24	1.10

a. Which fuel gave the greatest temperature rise in the experiment? .... [1]

b. Which fuel produced the most energy per gram? .... [1]

c. i. Which fuel would be the cheapest to transport per litre? Give a reason for your answer.

.... [2]

ii. Apart from your answer to part **c. i.** give one other factor which would influence how expensive it is to transport a fuel.

.... [1]

2. Complete these equations for the complete combustion of different compounds.

a. .... $H_2 + O_2 \rightarrow$ ....   [2]

b. .... $C_2H_6 +$ .... $\rightarrow$ .... $+$ ....   [2]

c. $C_7H_{16} +$ .... $\rightarrow$ .... $+$ ....   [2]

d. .... $H_2S +$ .... $\rightarrow$ .... $+$ .... SO_2 [2]

Language lab

Join up these fragments to form a sentence describing the transfer of charge in an electrochemical cell with zinc and copper electrodes.

..... zinc ions and .....   Zinc is more reactive .... at releasing electrons, forming .....

..... pole of the cell. ·   ..... than copper, so zinc is better .... becoming the negative .....

....

.... [2]

1. The diagram shows three electrochemical cells.

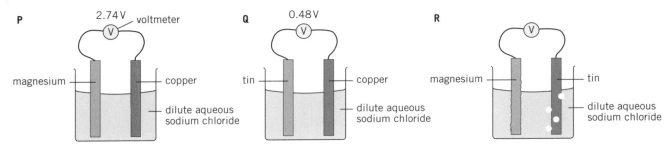

a. The voltage produced in an electrochemical cell is a measure of the reactivity of the two metals.

    i.   Look at cells **P** and **Q**. State which metal, magnesium or tin, is more reactive. Explain your answer.

         .... [2]

    ii. Deduce the voltage of cell **R**. .... [1]

b. Part of the reactivity series is shown below:

              lithium   magnesium   zinc   iron   tin   copper   silver

    most reactive ───⟶ least reactive

    i.   Which metal could replace magnesium in cell **P** to get a higher voltage?

         .... [1]

    ii. Which metal could replace copper in cell **Q** to get a higher voltage?

         .... [1]

    iii. Which reactive metal could replace magnesium in cell **R** to obtain a lower voltage?

         .... [1]

Language lab

Complete these sentences about fuel cells using words from the list.

alkali   electrons   external   negative   oxygen   platinum   porous   reactions

A fuel cell consists of two .... electrodes coated with .... The electrolyte

is either an acid or an .... Hydrogen and .... are bubbled through the

porous electrodes where the .... take place. When connected to an ....

circuit, ... flow from the .... electrode to the positive electrode.   [8]

1. The diagram shows a hydrogen–oxygen fuel cell.

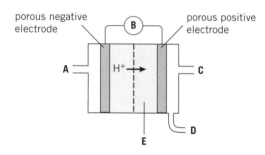

porous negative electrode

porous positive electrode

    a. Give the name of the instrument labelled **B**. .... [1]

    b. The product of the reaction in the fuel cell is water. Which letter (**A, C, D,** or **E**) shows where water is collected?

        .... [1]

    c. Which letter (**A, C, D,** or **E**) shows where hydrogen enters the cell? .... [1]

    d. On the diagram above draw an arrow to show the direction of the flow of electrons in the external circuit.   [1]

2. Complete these half-equations for the reactions taking place in the fuel cell.

    a. .... H$_2$ + ... OH$^-$ → .... H$_2$O + ...   [2]

    b. .... + ... H$_2$O + ... → .... OH$^-$ [2]

3. Give two advantages of fuel cells instead of a petrol engine to power a car.

    ....

    .... [2]

Language lab

Complete these sentences about methods for following the course of a reaction using words from the list.

decreases   products   quickly   rate   second   time   used   volume

To find the rate of reaction we can either measure how .... the reactants are

.... up or how quickly the .... are formed. To calculate the .... of

reaction we need to find out how some measurement changes with .... For

example, the .... of gas given off per ...., or how the mass of the reaction

mixture .... with time.   [8]

1. Put these in order of increasing rate of reaction.

    A Immediate precipitation   **B** Rusting   **C** Paint drying   **D** Magnesium burning in air

    .... [1]

2. The diagrams show the reaction of 50% nitric acid with copper.

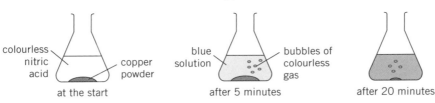

    colourless nitric acid — copper powder   at the start

    blue solution — bubbles of colourless gas   after 5 minutes

    after 20 minutes

    a. Give three pieces of information from the diagram that show a chemical reaction is occurring.

        ....

        .... [3]

    b. Suggest three different ways by which you could measure the rate of this reaction.

        1 ....

        2 ....

        3 .... [3]

3. Suggest why the course of the following reaction could be monitored by measuring electrical conductivity.

    $$H_2O_2(aq) + 2I^-(aq) + 2H^+(aq) \rightarrow 2H_2O(l) + I_2(aq)$$

    .... [2]

Language lab

Join up these fragments to form a sentence describing how the rate of reaction between zinc and hydrochloric acid changes with time.

..... per second is high, but ..... At the start of the reaction, the .... of the graph decreasing.

..... volume of hydrogen given off .... as the reaction proceeds, the volume .....

..... of hydrogen given off per second decreases, ... which is shown by the gradient .....

....

....

.... [2]

1. The decomposition of hydrogen peroxide is speeded up by catalysts.

$$2H_2O_2(aq) \rightarrow 2H_2O(l) + O_2(g)$$

A student investigated how the rate of reaction changes when two different catalysts are used.

The results using catalyst **A** are shown below.

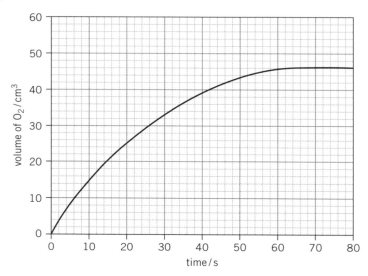

a. At what time is the reaction just complete? .... [1]

b. What volume of gas has been released when the reaction is just complete?

.... [1]

c. What volume of gas has been produced in the first 30 seconds of the reaction?

.... [1]

d. The reaction was repeated using catalyst **B**. The results are shown in the table.

time / s	0	4	10	20	30	40	50	60	70	80
volume / cm³	0	16	28	38	44	45	46	46	46	46

Plot a graph of these results on the same grid as for catalyst **A** above. Draw the curve of best fit through the points. [2]

Complete these sentences about catalysts using words from the list.

end   increases   less   rate   reaction   unchanged

A catalyst is a substance that .... the .... of a reaction. The catalyst

is .... at the .... of the reaction. A catalyst works by providing a

mechanism for the .... which needs .... energy.   [6]

1. a. Calculate the surface area of a cube of marble of dimensions 2 cm × 2 cm × 2 cm.
        (If you are not sure how to do this see unit 23.5.)

        .... [1]

    b. The cube is cut up into 8 cubes of dimensions 1 cm × 1 cm × 1 cm.
        Calculate the total surface area of all these cubes.

        .... [1]

    c. Which set of cubes will react faster with hydrochloric acid? Explain your answer.

        .... [2]

2. 4 g of large marble chips reacted with 20 cm³ of 0.5 mol/dm³ hydrochloric acid.

    $$CaCO_3(s) \quad + \quad 2HCl(aq) \quad \rightarrow \quad CaCl_2(aq) \quad + \quad CO_2(g) \quad + \quad H_2O(l)$$

The experiment was repeated with 4 g of medium-sized marble chips, then with 4 g of small marble chips. All other
conditions remained the same.

    a. On the axes below draw a sketch graph to show how the volume of carbon dioxide released changes with time
        using large, **L**, medium, **M**, and small, **S**, marble chips. Label your lines **L**, **M**, and **S**.

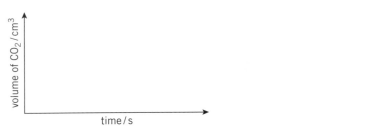

        [2]

    b. The experiment was repeated but this
        time measuring the decrease in the mass
        of the reaction mixture with time. In the
        space on the right, draw a sketch graph
        to show how the mass of the reaction
        mixture changes with time for the small
        and the large marble chips.

        [4]

Language lab

Complete these sentences about the collision theory using words from the list.

bonds   collide   energy   frequency   increases   rate

In order to react, particles must .... with each other. The collisions must have

enough .... to break .... to allow a reaction to happen. Increasing

the concentration of a reactant .... the .... of collisions and so

increases the .... of reaction. [6]

1. Use the particle diagrams below to answer the following questions about the reaction:

$$Mg(s) \quad + \quad 2HCl(aq) \quad \rightarrow \quad MgCl_2(aq) \quad + \quad H_2(g)$$

a. Complete the diagram on the right to show the particles of metal, acid and water in a concentrated solution of acid.

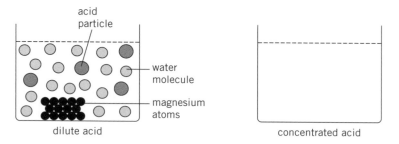

dilute acid   concentrated acid   [3]

b. Complete the diagram on the right to show the relative number of acid, magnesium, and water particles when the reaction is nearly complete.

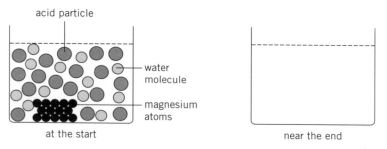

at the start   near the end   [3]

2. Explain why the rate of the reaction shown below increases when the pressure increases.

$$2SO_2(g) \quad + \quad O_2(g) \quad \rightarrow \quad 2SO_3(g)$$

....

.... [2]

Language lab

Link the words **A** to **D** on the left with the phrases **1** to **4** on the right.

| A Activation energy | 1 The number of collisions in a given time |

| B Effective collisions | 2 The energy associated with moving particles |

| C Frequency of collisions | 3 The minimum energy particles must have when they collide in order to react |

| D Kinetic energy | 4 Collisions which result in a reaction taking place |

[2]

1 A student investigated how increasing the temperature affects the rate of the reaction of magnesium with 1.0 mol/dm³ hydrochloric acid. The student measured the volume of hydrogen given off in 1 minute at seven different temperatures. The table shows the results.

temperature / °C	20	30	41	45	50	55	60
volume of H_2 / cm³	10	20	44	56	80	110	160

a. Plot a graph of these results on the grid below. Draw the best-fit curve through the points.

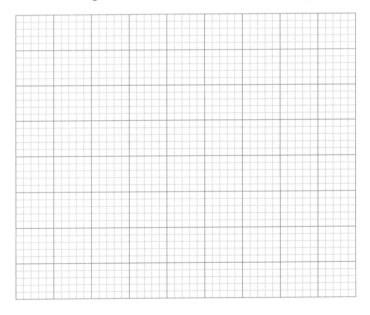

[5]

b. Describe the effect of temperature on the rate of reaction.

....

.... [2]

Language lab

Use the clues below to do this crossword.

Across: 5. Process of producing images on film

6. Metal formed when light strikes black and white film

Down: 1. Oxidation–reduction reaction

2. Green pigment in plants

3. Type of radiation that helps us see

4. Ion which gets oxidised in black and white photography

[6]

1. **a.** Complete the chemical equation for photosynthesis.

$$\text{.............................} \ + \ \text{.............................} \ \rightarrow \ C_6H_{12}O_6 \ + \ 6O_2 \qquad [2]$$

b. The graph shows how the rate of photosynthesis is affected by the brightness (intensity) of light.

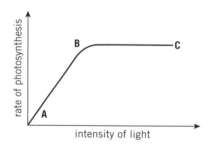

i. Describe how rate of photosynthesis varies with light intensity from **A** to **B**.

.... [1]

ii. Suggest why the graph is horizontal between **B** and **C**.

.... [1]

2. The equation shows the photochemical reaction taking place when a black and white film is exposed to light.

$$2AgBr \ \rightarrow \ 2Ag \ + \ Br_2$$

Complete the two half-equations for this reaction and state whether oxidation or reduction is taking place.

................Ag^+ + ... $\rightarrow$ 2Ag   Oxidation or reduction? ....

$2Br^-$ $\rightarrow$ .... + ...   Oxidation or reduction? .... [3]

Language lab

Search for the names of six words about reversible reactions. Words may go downwards, upwards, forwards, or backwards, but not diagonally. Write the names in the space below.

P	I	L	L	O	W	P	X	E	S
A	N	C	H	E	W	O	P	T	I
R	N	D	T	E	A	S	I	O	M
R	E	V	E	R	S	I	B	L	E
T	N	Y	E	C	X	T	O	E	T
T	D	Y	N	A	M	I	C	L	S
A	N	H	Y	D	R	O	U	S	Y
Z	Z	F	E	D	I	N	O	U	S
H	Y	D	R	A	T	E	D	X	X

[6]

1. Hydrated and anhydrous cobalt(II) chloride can be converted to one another.

$$CoCl_2.6H_2O \rightleftharpoons CoCl_2 + 6H_2O \qquad \Delta H \text{ is } +$$

pink hydrated   blue anhydrous

a. What is the meaning of the symbol $\rightleftharpoons$? .... [1]

b. How could you change pink cobalt(II) chloride to blue cobalt(II) chloride?

.... [1]

c. Explain why the reaction blue cobalt(II) chloride to pink cobalt(II) chloride is exothermic.

....

.... [2]

2. When a mixture of hydrogen and iodine is heated, an equilibrium mixture with hydrogen iodide is formed.

$$H_2(g) + I_2(g) \rightleftharpoons 2HI(g)$$

In the box on the right, draw the molecules in this equilibrium mixture, which contains more product than reactants.

Use ● to represent a molecule of I_2

Use ○ to represent a molecule of H_2

Use □ to represent a molecule of HI

[3]

Complete these sentences about the effect of temperature on equilibrium using words from the list.

endothermic   favours   heat   reverse   shifts   temperature

If a reaction is exothermic in the forward direction, it will be .... in the reverse

direction. For an exothermic reaction, when .... increases, the equilibrium

.... in the direction of the .... reaction. It .... the

endothermic change where .... is taken in.   [6]

1. Sulfur dioxide reacts with oxygen to form an equilibrium mixture with sulfur trioxide.

$$2SO_2(g) + O_2(g) \rightleftharpoons 2SO_3(g) \quad \text{energy released}$$

   Complete the following sentences about this reaction.

   a. When oxygen is removed the position of equilibrium shifts to the ....   [1]

   b. Decreasing the pressure shifts the position of equilibrium to the ....

      because there are .... moles of gas molecules in the equation on the

      ....   [3]

2. In which direction does the position of equilibrium shift when the pressure on each of these reactions is increased?

   a. $CO(g) + 2H_2(g) \rightleftharpoons CH_3OH(g)$ ....   [1]

   b. $CaCO_3(s) \rightleftharpoons CaO(s) + CO_2(g)$ ....   [1]

   c. $4HCl(g) + O_2(g) \rightleftharpoons 2H_2O(g) + 2Cl_2(g)$ ....   [1]

   d. $2HCl(g) \rightleftharpoons H_2(g) + Cl_2(g)$ ....   [1]

3. When bismuth trichloride, $BiCl_3$, is added to water the following reaction occurs.

$$BiCl_3(aq) \quad + H_2O(l) \rightleftharpoons \quad BiClO(s) \quad + 2HCl(aq)$$
   colourless solution   white precipitate

   a. What would you see when concentrated HCl is added to the mixture?

      .... [1]

   b. What would you see when a large volume of water was added to the reaction mixture?

      .... [1]

Language lab

Use the clues below to do this crossword.

Across: 3. Oxidation and reduction together
5. Oxidation .... tells us how oxidised or reduced a substance is
6. Gain of electrons
7. A good reducing agent in Group IV

Down: 1. Oxidation is .... of electrons
2. A substance that removes electrons from another substance
4. A negative particle that is transferred in redox reactions   [7]

1. Draw arrows to show which of the elements or compounds have undergone oxidation and which have undergone reduction. An example is given below.

Example:

oxidation

$$CuO(s) + H_2(g) \longrightarrow Cu(s) + H_2O(l)$$

reduction

a. $2H_2(g) + O_2(g) \rightarrow 2H_2O(l)$ [2]

b. $PbO(s) + H_2(g) \rightarrow Pb(s) + H_2O(l)$ [2]

c. $Fe_2O_3(s) + 3C(g) \rightarrow 2Fe(s) + 3CO(g)$ [2]

d. $C(s) + H_2O(g) \rightarrow CO + H_2(g)$ [2]

e. $ZnO(s) + C(s) \rightarrow CO(g) + Zn(s)$ [2]

f. $3Fe(s) + 4H_2O(g) \rightarrow Fe_3O_4(s) + 4H_2(g)$ [2]

Link the words **A** to **D** on the left with descriptions **1** to **4** on the right.

A Oxidation	**1** This happens when an atom gains one or more electrons
B Oxidising agent	**2** A substance that gives electrons to another substance
C Reducing agent	**3** This happens when an atom loses one or more electrons
D Reduction	**4** A substance that removes electrons from another substance

[2]

1. Identify the oxidising and reducing agents in these equations.
   Underline the reducing agent and put a ring around the oxidising agent.

   a. $2Mg + O_2 \rightarrow 2MgO$ 　　　　　　　　　　　　　　　[1]

   b. $PbO + H_2 \rightarrow Pb + H_2O$ 　　　　　　　　　　　　　[1]

   c. $2I^- + Cl_2 \rightarrow I_2 + 2Cl^-$ 　　　　　　　　　　　　[1]

   d. $H_2O_2 + 2I^- + 2H^+ \rightarrow I_2 + 2H_2O$ 　　　　　　[1]

2. Complete these half-equations. Balance the charges by adding electrons.
   State whether oxidation or reduction has taken place.

   a. $Ca \rightarrow Ca^{2+} +$ .... 　　　Oxidation or reduction? .... [2]

   b. $Cl_2 +$ .... $\rightarrow$ .... Cl^- 　　　Oxidation or reduction? .... [2]

   c. $Al^{3+} +$ .... $\rightarrow$ .... 　　　Oxidation or reduction? .... [2]

   d. $Fe^{2+} \rightarrow Fe^{3+} +$ .... 　　　Oxidation or reduction? .... [2]

   e. $O_2 +$ .... $\rightarrow$ .... O^{2-} 　　　Oxidation or reduction? .... [2]

   f. $Pb^{4+} + 2e^- \rightarrow$ .... 　　　Oxidation or reduction? .... [2]

   g. .... $Br^- \rightarrow$ .... $+$ .... 　　　Oxidation or reduction? .... [2]

3. Describe the colour change when potassium manganate(VII) oxidises excess colourless sodium sulfite to colourless sodium sulfate.

   From .... to .... [2]

Use the clues below to do this crossword.

Across: 1. Element used to make a common acid

2. A pH above 7 is ....

5. An indicator used to test if a solution is acidic or alkaline

7. A substance which changes colour at a particular pH

Down: 1. A slightly alkaline substance we use to clean ourselves

2. A substance with a pH below 7

3. A substance with pH 7 is ....

4. Uni .... al Indicator is used to find the pH of a solution

6. A pH value less than two   [9]

1. Link the phrases **A** to **E** on the left with the pH values **1** to **5** on the right.

A highly acidic		**1** pH 6
B neutral		**2** pH 8
C highly alkaline		**3** pH 1
D weakly acidic		**4** pH 14
E weakly alkaline		**5** pH 7

[3]

2. **a.** Link these acids and alkalis by choosing the correct formula from the list.

$Ca(OH)_2$ CH_3COOH H_2CO_3 HNO_3 H_3PO_4 H_2SO_4 $NaOH$ NH_3

i. ammonia .... [1]   **ii.** ethanoic acid .... [1]

iii. carbonic acid .... [1]   **iv.** sulfuric acid .... [1]

v. calcium hydroxide .... [1]   **vi.** phosphoric acid .... [1]

vii. nitric acid .... [1]   **viii.** sodium hydroxide .... [1]

b. What group of atoms is found in all alkalis?

.... [1]

3. Many acids and alkalis are corrosive. What dos the word *corrosive* mean?

.... [1]

Complete these sentences about acids using words from the list.

        atoms   dioxide   hydrogen   hydroxides   replaced   salt   water

Acids react with many metals to form a salt and .... A salt is a substance formed

when particular hydrogen .... in an acid are .... by a metal. Acids also

react with some metal .... to produce a .... and water, and with carbonates

to produce a salt, ..., and carbon ....   [7]

1. Link the colours of these indicators to the following pH values.

                    pH 3   pH 7   pH 9   pH 13

    a. Litmus is blue at pH ....   [1]

    b. Methyl orange is red at pH ....   [1]

    c. Universal indicator is green at pH ....   [1]

    d. Litmus is red at pH ....   [1]

    e. Methyl orange is yellow at pH ....   [1]

2. Complete these word equations.

    a. zinc oxide + hydrochloric acid → ....   [1]

    b. iron + sulfuric acid → ....   [1]

    c. sulfuric acid + lead carbonate → ....   [1]

    d. hydrochloric acid + tin oxide → ....   [1]

3. Complete the balanced chemical equations for these reactions.

    a. $Zn + H_2SO_4 →$ ....   [1]

    b. $MgO + $ ....$HNO_3 →$ ....   [2]

    c. $CuCO_3 + $ ....$HCl →$ ....   [2]

    d. $Na_2CO_3 + $ ....$HCl →$ ....   [2]

    e. $Ca + $ ....$HCl →$ ....   [2]

    f. ....$KOH + H_2SO_4 →$ ....   [2]

Complete these sentences about bases using words from the list.

acid　　alkalis　　oxides　　salt　　soluble　　sulfate　　sulfuric

A base is a substance which reacts with an .... to form a salt. Bases that are

.... in water are called .... Many metal .... are basic because

they react with acids to form a .... and water. Ammonia reacts with ....

acid to form ammonium ...., so ammonia is also a base. [7]

1. Complete these word equations.

   a. sodium hydroxide + nitric acid → .... [1]

   b. calcium oxide + hydrochloric acid → .... [1]

   c. barium hydroxide + nitric acid → .... [1]

   d. hydrochloric acid + barium oxide → .... [1]

2. Complete the balanced chemical equations for these reactions.

   a. $Zn + H_2SO_4 →$ .... [1]

   b. $MgO + ... HNO_3 →$ .... [2]

   c. $........ NaOH + H_2SO_4 →$ .... [2]

   d. $Ca(OH)_2 + ... HNO_3 →$ .... [2]

   e. $........ NH_3 + H_2SO_4 →$ .... [2]

3. Complete the equation below for the reaction of an alkali with an ammonium salt.

   $$Ca(OH)_2 + ... NH_4Cl → CaCl_2 + ... H_2O + ...$$ [2]

4. Explain why farmers may add calcium oxide or calcium carbonate to the soil.

   ....

   .... [2]

5. Ammonia reacts with water: $NH_3(g) + H_2O(l) \rightleftharpoons NH_4^+(aq) + OH^-(aq)$

   How does this equation show that aqueous ammonia is an alkali?

   .... [1]

Language lab

Complete these sentences about strong and weak acids using words from the list.

| all | hydrogen | ions | ionised | molecules | partially | water |

Aqueous solutions of acids contain .... ions. In strong acids .... the

acid .... are dissociated (.....................) to form hydrogen .... and

anions. When weak acids dissolve in .... they become ....

dissociated. [7]

1. Use the information in the table to answer parts **a.** and **b.**

acid 0.1 mol/dm³ solution	relative electrical conductivity	pH	rate of reaction
ethanoic, CH_3COOH	0.5		
hydrochloric, HCl	25		
methanoic, HCOOH	2	2.4	
sulfuric, H_2SO_4	40		

   a. Hydrogen ions conduct electricity better than other ions. The greater the concentration of hydrogen ions, the better is the electrical conductivity.
      Complete the third column of the table using the following pH values: pH 0.7, pH 1.0, pH 2.9. [3]

   b. The rate of reaction of magnesium with different acids depends on the hydrogen ion concentration in the acid. Suggest whether the reaction of magnesium with each of these acids is fast or slow. Write your answers in the fourth column. [4]

2. The equation below shows the ions present in the reactants and products of a neutralisation reaction.

$$H^+(aq) + NO_3^-(aq) + Na^+(aq) + OH^-(aq) \rightarrow NO_3^-(aq) + Na^+(aq) + H_2O(l)$$

   a. Cancel out the spectator ions in this equation. [1]

   b. Write the ionic equation for this reaction.

   .... [1]

   c. Explain, in terms of ions, why this is a neutralisation reaction.

   .... [1]

1. Link the oxides **A** to **D** on the left with the descriptions **1** to **4** on the right.

A Acidic	**1** Oxides that react with acids but not with alkalis
B Amphoteric	**2** Oxides that react with alkalis but not with acids
C Basic	**3** Oxides that react with neither acids nor bases
D Neutral	**4** Oxides that react with both acids and alkalis

[2]

1. Complete these equations to show the reactions of some oxides with either acids or bases.

   a. $MgO + \ldots\ldots HCl \rightarrow \ldots\ldots\ldots\ldots + \ldots\ldots\ldots\ldots$ 　　　　　　　　[2]

   b. $SO_2 + \ldots\ldots NaOH \rightarrow Na_2SO_3 + \ldots\ldots\ldots\ldots$ 　　　　　　　　[2]

   c. $CuO + H_2SO_4 \rightarrow \ldots\ldots\ldots\ldots\ldots + \ldots\ldots\ldots\ldots$ 　　　　　　　　[1]

   d. $CO_2 + \ldots\ldots NaOH \rightarrow Na_2CO_3 + \ldots\ldots\ldots\ldots$ 　　　　　　　　[2]

   e. $ZnO + \ldots\ldots HNO_3 \rightarrow \ldots\ldots\ldots\ldots + \ldots\ldots\ldots\ldots$ 　　　　　　　　[2]

   f. $CaO + H_2SO_4 \rightarrow \ldots\ldots\ldots\ldots\ldots + \ldots\ldots\ldots\ldots$ 　　　　　　　　[1]

2. Complete these equations to show the reactions of some oxides with water to form acids or alkalis.

   a. $SO_2 + H_2O \rightarrow \ldots\ldots\ldots\ldots$ 　　　　　　　　[1]

   b. $CO_2 + H_2O \rightarrow \ldots\ldots\ldots\ldots$ 　　　　　　　　[1]

   c. $CaO + H_2O \rightarrow \ldots\ldots\ldots\ldots$ 　　　　　　　　[1]

   d. $P_4O_6 + \ldots\ldots H_2O \rightarrow \ldots\ldots H_3PO_3$ 　　　　　　　　[2]

   e. $Na_2O + H_2O \rightarrow \ldots\ldots\ldots\ldots$ 　　　　　　　　[2]

3. Complete the equation for the reaction of zinc oxide with potassium hydroxide.

   $ZnO + 2KOH \rightarrow \ldots\ldots\ldots\ldots\ldots + \ldots\ldots\ldots\ldots\ldots$ 　　　　　　　　[2]

Language lab

Search for the names of seven words about making a salt from a metal oxide. Names of pieces of apparatus are included. Words go forwards or downwards only. Write the names in the space below.

[7]

C	R	Y	S	T	A	L	X	A	S
E	N	C	R	I	P	Z	I	O	N
F	I	L	T	R	A	T	E	X	P
O	N	P	R	E	S	T	A	D	I
H	S	E	Z	F	I	L	T	E	R
E	O	X	I	D	E	Y	I	P	X
A	L	L	D	S	L	Y	M	O	L
T	U	P	T	P	P	R	E	D	P
X	B	I	M	Y	L	L	O	D	L
T	L	U	R	E	W	H	E	L	K
K	E	V	A	P	O	R	A	T	E

1. Zinc sulfate can be made by first warming sulfuric acid with excess zinc.

   a. How is a solution of zinc sulfate obtained from the reaction mixture?

   .... [1]

   b. The solution of zinc sulfate is crystallised. Describe how you could obtain pure dry crystals of zinc sulfate from a mixture of the crystals and remaining solution.

   ....

   ....

   .... [3]

2. Crystals of copper(II) sulfate can be made by warming excess copper(II) carbonate with sulfuric acid. Put the following stages in the correct order.

   A   Allow the solution to cool and deposit crystals.

   B   Pour the filtrate into an evaporating basin.

   C   Wash and dry the crystals.

   D   Filter the mixture to remove excess copper(II) carbonate.

   E   Warm the filtrate until the solution is very concentrated.

   F   Filter off the crystals.

   The order is .... [2]

3. Write symbol equations for the formation of these salts using suitable acids.

   a. Calcium sulfate from calcium oxide.

   .... [2]

   b. Zinc chloride from zinc.

   .... [2]

Use the clues below to do this crossword.

Across:
1. The amount added from the burette in a titration
3. The solid dissolved to form a solution
6. This changes colour at the neutralisation point of a titration
7. This neutralises an alkali
8. 36 g of NaOH (M_r 40) is 0.X moles NaOH
9. The .... -point of a titration is when the indicator changes colour.

Down:
2. Procedure for finding the concentration of an alkali by adding acid from a burette
4. Long glass tube used in titrations
5. This is used to deliver an exact volume of solution [9]

1. The diagram shows the stages in making a soluble salt (sodium chloride) by neutralising an alkali (sodium hydroxide) with an acid (hydrochloric acid). The first three stages are shown in the diagram.

indicator

Phenol-phthalein

indicator turns pink

sodium hydroxide solution

acid added from burette

solution is still pink

on adding one more drop, pink colour suddenly disappears

Describe the next three stages of the procedure to get colourless crystals of sodium chloride.

....

....

.... [3]

2. Write symbol equations for the formation of these salts using suitable acids.

a. Sodium nitrate from sodium hydroxide.

.... [2]

b. Ammonium sulfate from ammonia.

.... [2]

Language lab

Complete these sentences about solubility using words from the list.

ammonium   carbonates   compounds   hydroxides   nitrates   precipitate   solutions

Salts such as ...., sodium salts, and .... salts are soluble in water.

Many .... and .... are insoluble, except those from Group I. An

insoluble substance formed when two .... of soluble .... are mixed

is called a ....   [7]

1. Are the following compounds soluble or insoluble? Write 'soluble' or 'insoluble' in the spaces provided.

    a. sodium hydroxide .... [1]   b. potassium nitrate .... [1]

    c. ammonium chloride .... [1]   d. lead chloride .... [1]

    e. iron(II) hydroxide .... [1]   f. barium sulfate .... [1]

2. Crystals of lead iodide can be made from solutions of lead nitrate and potassium iodide. Put the following stages in the correct order.

    A Filter the mixture.

    B Make up aqueous solutions of lead nitrate and potassium iodide.

    C Dry the residue of lead iodide in a warm oven.

    D Rinse the residue on the filter paper with distilled water.

    E Mix the solutions. A yellow precipitate forms.

    The order is .... [2]

3. a. The equation below shows the ions present in the reactants and products of a precipitation reaction.

        $Ag^+(aq) + NO_3^-(aq) + K^+(aq) + Br^-(aq) \rightarrow AgBr(s) + NO_3^-(aq) + K^+(aq)$

        i. Cancel out the spectator ions in this equation.   [1]

        ii. Write the ionic equation for this reaction.

             .... [1]

    b. Write an ionic equation for this reaction.

        $BaCl_2(aq) + K_2SO_4(aq) \rightarrow BaSO_4(s) + 2KCl(aq)$

        .... [2]

Language lab

Complete these sentences about collecting gases using words from the list.

air   collect   displacement   downward   less   upward

When a gas is denser than ...., you collect it by .... displacement of air.

If a gas is .... dense than air, you .... it by ....

.... of air.   [6]

1. The diagram shows four ways of collecting gases in the laboratory.

A   B   C   D

measuring cylinder

Which method of gas collection, **A**, **B**, **C**, or **D**, is used for:

a. Measuring the volume of a gas that is sparingly soluble in water? .... [1]

b. Collecting a gas that is lighter than air? .... [1]

c. Collecting a gas that is heavier than air? .... [1]

2. Which method of gas collection, **A**, **B**, **C**, or **D**, involves:

a. Upward displacement of air? .... [1]

b. Collection in a gas syringe? .... [1]

3. Link the gases **A** to **D** on the left with the best test results **1** to **4** on the right.

A Ammonia		1 Bleaches damp litmus paper
B Carbon dioxide		2 Turns damp red litmus paper blue
C Chlorine		3 Turns acidified aqueous potassium manganate(VII) colourless
D Sulfur dioxide		4 Turns limewater milky

[2]

Complete these sentences about flame tests using words from the list.

coloured   edge   flame   lithium   luminous   potassium   sample   wire

A .... is put on the end of a platinum .... and placed at the .... of a

non-............................ Bunsen .... If .... is present, the flame is

.... red. If .... is present, the flame has a lilac colour.   [8]

1. Complete the table showing what happens when an aqueous solution of ions reacts with aqueous sodium hydroxide
   and aqueous ammonia.

aqueous ion	reaction with sodium hydroxide	reaction with aqueous ammonia
$Al^{3+}(aq)$	at first .... in excess	at first .... in excess ....
$Cr^{3+}(aq)$	at first .... in excess	at first .... in excess ....
$Cu^{2+}(aq)$	at first .... in excess	at first .... in excess ....
$Fe^{3+}(aq)$	at first .... in excess	at first .... in excess ....

[16]

Complete these sentences about the test for halides using words from the list.

chloride   cream   drops   nitrate   precipitate   solution

A few .... of nitric acid are added to the .... thought to be a halide.

Aqueous silver .... is then added. If a .... is present, a white

.... is seen. If a bromide is present, a ....-coloured precipitate is seen.   [6]

1. Link the anions **A** to **D** on the left with the test results **1** to **4** on the right.

A Carbonate ions	**1** Ammonia produced when heated with sodium hydroxide and aluminium foil
B Aqueous nitrate ions	**2** White precipitate formed on addition of aqueous nitric acid and barium nitrate
C Aqueous sulfite ions	**3** SO_2 produced when warmed with dilute hydrochloric acid
D Aqueous sulfate ions	**4** Effervescence of carbon dioxide on the addition of an acid

[2]

2. a. Write a symbol equation for the reaction of aqueous silver nitrate with aqueous sodium chloride. Include state symbols.

   .... [2]

   b. Convert this equation into an ionic equation.

   .... [1]

3. a. Write a symbol equation for the reaction of aqueous barium chloride with aqueous sodium sulfate. Include state symbols.

   .... [3]

   b. Convert this equation into an ionic equation.

   .... [1]

Language lab

Use the clues below to do this crossword.

Across: 4. Symbol for metal in Group II that is more reactive than calcium

5. Green gas

6. Element in Group III having lowest A_r

8. A transition element with fewer electrons than copper

Down: 1. Metal in Group II and Period 3

2. Symbol for argon

3. Its ions give yellow colour to a flame

4. Yellow element in Group VI

5. This element is found as graphite

7. Symbol for element that has an amphoteric oxide

[10]

1. The table shows some information about some of the elements in Period 3.

element	Na	Mg	Al	Si	P	S	Cl
electronic structure	2,8,1						
melting point / °C	98	649	660	1410	590	119	−101
formula of typical compounds	NaCl Na$_2$O	MgCl$_2$ MgO	AlCl$_3$	SiCl$_4$ SiH$_4$	PCl$_3$ PH$_3$	H$_2$S	HCl

a. Complete the 2nd and 4th lines of the table. [4]

b. i. Describe how the melting points of the elements change across the Period.

.... [2]

ii. What type of structures are Na, Mg, and Al?

.... [2]

iii. Explain in terms of structure and bonding why Si has the highest melting point in this Period.

....

.... [2]

iv. Explain in terms of structure and bonding why the melting points of P, S, and Cl are relatively low.

....

.... [2]

Complete the sentences about Group I elements using words from the list.

alkali   densities   hydrogen   hydroxide   melting

The .... metals have low .... points and .... compared with

most other metals. They react with water to form a metal .... and .... [5]

1. The table shows some properties of some Group I metals.

Group I metal	density / g/cm^3	melting point / °C	metallic radius / nm	observations when the metal reacts with water
lithium	0.53	181	0.157	Moves over the surface very slowly Fizzes gently Does not melt or go into a ball Does not burst into flame
sodium	0.97	98	0.191	....
potassium	0.86	....	0.235	Moves over surface very rapidly Fizzes very rapidly Melts and goes into a ball then bursts into flame Slight 'pop' when reaction near the end
rubidium	1.53	39	....	....

a. Complete the table.   [8]

b. Caesium is below rubidium in the Periodic Table. Predict a value for the density of caesium.

.... [1]

Language lab

Complete these sentences about the displacement reactions of halogens using words from the list.

bromine   chlorine   colourless   halide   halogen   less   more   orange

When aqueous .... is added to a .... solution of potassium

bromide, the solution turns .... because .... has been displaced.

This is because a .... reactive .... displaces a .... reactive

halogen from an aqueous solution of its ....   [8]

1. The table shows some properties of fluorine, chlorine, bromine, and iodine.

halogen	melting point / °C	boiling point / °C	state at −40 °C	depth of colour	atomic radius / nm
fluorine	−220	−188			
chlorine	−101	−35			
bromine	−7	59			
iodine	114	184			

a. What is the trend in the melting points of the halogens?

.... [1]

b. Use the values of the melting and boiling points in the table to deduce the state of the halogens at −40 °C. Write your answers in the table.   [4]

c. Draw an arrow in the 5th column to show the trend in the depth of colour (light → dark).   [1]

d. Draw an arrow in the 6th column to show the trend in atomic radius (smaller → larger).   [1]

2. What would you observe when an aqueous solution of bromine is added to an aqueous solution of potassium iodide? Explain these observations.

....

....

.... [4]

Language lab

Complete these sentences about the noble gases using words from the list.

atoms   eight   electron   gaining   more   noble   sharing   shell   stable

The Group VIII gases (............................ gases) are unreactive because their ....

arrangements make them .... It is difficult for their .... to form ionic bonds by

.... or losing electrons, or covalent bonds by .... electrons. Helium is

particularly unreactive because there cannot be .... than two electrons in the first

.... The other gases have .... electrons in their outer shell, which is a stable

electronic structure.   [9]

1. Link the Group VIII gases **A** to **D** on the left with their uses **1** to **4** on the right.

A argon		1 Produces a red glow for advertising signs
B helium		2 Used in car headlamps and for lasers
C neon		3 Filling balloons and airships
D krypton		4 To provide an inert atmosphere in welding

[2]

2. The density of air and some noble gases at r.t.p. in g/dm³ are given below.

Air 1.20   Ar 1.78   He 0.18   Kr 3.74   Ne 0.90

Which gases could you use to fill balloons that float upwards in air?

.... [1]

3. a. Use ideas about electron arrangement to suggest why the Group VIII elements are monatomic and not diatomic.

....
.... [2]

b. If we fire high-speed electrons at an atom, they can knock an electron out of an atom.

Complete the equation which represents this reaction.

$$Ar \rightarrow Ar^+ + \text{..................}$$ [1]

Language lab

Use the clues below to do this crossword.

Across: 2. First three letters of atom with 27 electrons
4. A property of transition element ions
6. Abbreviation for oxidation number?
8. Iron, cobalt, and nickel are magn ....
9. Symbol for nickel
10. Property of transition element compound that speeds up a reaction

Down: 1. Metal used for making bridges and cars
2. .... plating on bicycles and taps
3. Iron is hard, sodium is ....
4. Ions of this element give a dark blue solution with excess aqueous ammonia
5. Transition elements have a very high ....
7. Copper has a high melting ....

[12]

1. The boxes below show some properties of a non-transition element and of a transition element. The boxes are muddled up. (M = metal)

A   Melting point 1890 °C

B   Forms a chloride of formula MCl_2 only

C   Density 3.51 g/cm^3

D   Forms chlorides that are pink and green

E   Forms chloride of type MCl_2, MCl_3, and MCl_4

F   Forms a colourless chloride

G   A compound of M is a good catalyst

H   Density 5.96 g/cm^3

I   Melting point 725 °C

J   Compounds of M show no catalytic activity

a. Which letters represent the properties of transition elements?

.... [3]

b. Give two other typical properties of transition elements which are not mentioned above.

.... [2]

2. Write the formulae of the transition element ions in the following compounds.

a. Ag_2O ....   b. $CuSO_4$ ....   [1]

c. $Cr(NO_3)_3$ ....   d. $Fe_2(SO_4)_3$ ....   [1]

3. Write an ionic equation, including state symbols, for the reaction of aqueous iron(III) ions with aqueous hydroxide ions.

.... [3]

Complete the sentences about alloys using words from the list.

alloyed   arrangement   difference   force   harder   mixtures   regular   prevents

Alloys are .... of metals or mixtures of metals with non-metals. Alloys are often

.... and stronger than pure metals. When a metal is .... with another metal,

the .... in the sizes of the metal atoms makes the .... of the layers in

the lattice less .... This .... the layers from sliding over each other so

easily when a .... is applied.   [8]

1. Complete the table about the uses and properties of different metals and alloys.

metal or alloy	use	properties which makes it suitable for the use
aluminium alloy (90.25% Al, 6% Zn, 3.75% Mg/Cu)	aircraft body	1. .... [1] 2. .... [1]
brass (70% Cu, 30% Zn)	door handles	1. .... [1] 2. .... [1]
bronze (95% Cu, 5% Sn)	.... [1]	1. .... [1] 2. Resistant to corrosion
cobalt alloy (65% Co, 30% Cr, 5% Mo)	1. .... [1] 2. .... [1]	1. Remains hard at high temperatures 2. Does not change shape easily

2. Suggest why an alloy of tin and lead has lower melting point than either pure tin or pure lead.

....

.... [2]

Language lab

Complete these sentences about the reactivity of metals using words from the list.

alkaline   blue   cold   hydrogen   lower   oxide   oxygen

The products formed by metals that react with .... water are a metal

hydroxide and .... The hydroxides are .... and so turn red litmus

.... The products formed by metals that only react with steam are a metal

.... and hydrogen. Copper does not react with water because it is ....

in the reactivity series than hydrogen and cannot take the .... away from the

hydrogen in the water.   [7]

1. Complete these equations for the reactions of metals with water or steam.

    a. .... Na(s)   + ... $H_2O(l)$ $\rightarrow$ ....   + ...   [3]

    b. .... Fe(s)   + ... $\rightarrow$ .... $Fe_3O_4(s)$ + ...   [3]

2. Some observations for the reactions of metals with water are given in the table.

metal	observations
barium	
calcium	Gives off bubbles rapidly with cold water, disappears quite quickly
lead	
magnesium	Gives a few bubbles with hot water, disappears slowly
zinc	Reacts when heated to red-heat with steam

    a. Put calcium, magnesium, and zinc in order of their reactivity. Put the most reactive first.

        .... [1]

    b. Barium is more reactive than calcium and lead is less reactive than zinc.
        Write the observations for barium and lead in the table above.   [3]

3. Put these metals in order of their reactivity using the information below.

metal	concentration of HCl(aq) / mol/dm³	observations
iron	0.5	slow bubbling
lead	6.0	slow bubbling
lithium	0.5	rapid stream of bubbles
magnesium	0.5	steady stream of bubbles

    least reactive .... most reactive   [1]

Language lab

Complete these sentences about the reactivity of metals using words from the list.

aqueous   atoms   displaces   higher   losing   solution

Calcium is .... in the reactivity series than copper. So calcium atoms are better at

.... electrons than copper. When calcium is added to .... copper(II)

sulfate, calcium .... the copper and copper metal is formed. Calcium ....

are converted to calcium ions, which go into ....   [6]

1. The diagram shows different metals placed in solutions of metal salts.

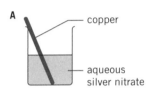

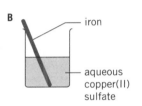

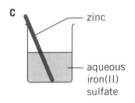

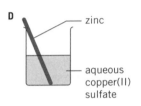

Some of the results are shown in the table.

experiment	colour at the start		colour after 20 minutes	
A	metal	brown	metal	silver-grey surface
	solution	colourless	solution	blue
B	metal	silvery grey	metal	
	solution		solution	
C	metal	grey	metal	
	solution	light green	solution	colourless
D	metal		metal	
	solution		solution	

a. Complete the table to show the colour changes.   [8]

b. Use the results to put the metals in order of their reactivity.

   least reactive .... most reactive   [1]

c. Explain why there would be no colour change when a copper rod is placed in aqueous zinc sulfate.

   .... [1]

Complete these sentences about the reactivity of aluminium using words from the list.

acids   freshly   layer   oxide   oxygen   reactivity   sticks   unreactive

Although aluminium is high in the .... series, samples of the metal exposed to the air do not

appear to react with water or dilute .... This is because .... made aluminium

reacts with .... in the air to form a thin .... of aluminium ....

on its surface, which is relatively .... The oxide layer .... to the metal surface

strongly so is not easily removed.   [8]

1. Identify the reducing agent and the oxidising agent in each of these equations.

   a. Fe(s) + CuO(s) $\rightarrow$ FeO(s) + Cu(s)

      reducing agent .... oxidising agent ....   [1]

   b. Fe_2O_3(s) + 3Mg(s) $\rightarrow$ 2Fe(s) + 3MgO(s)

      reducing agent .... oxidising agent ....   [1]

2. Manganese is extracted by reduction of manganese oxide with hot aluminium.
   Write a word equation for this reaction.

   .... [1]

3. Complete these equations for the reduction of some metal oxides.

   a. SnO_2 + ... C $\rightarrow$ .... + ...   [2]

   b. .... NiO + CO + H_2 $\rightarrow$ .... + ... + H_2O [2]

   c. .... PbO + C $\rightarrow$ .... + ...   [2]

4. The table shows the ease of reduction of some metal oxides.

metal oxide	ease of reduction
chromium(III) oxide	Reduced by carbon at 1200 °C
copper(II) oxide	Reduced by carbon below 400 °C
manganese(II) oxide	Reduced by carbon at 1400 °C
tin(IV) oxide	Reduced by carbon at 400 °C
silver(I) oxide	Reduced by heating alone at low temperatures

   a. Put the metals in order of their reactivity.

      least reactive .... most reactive   [2]

   b. Construct a balanced equation for the reduction of chromium(III) oxide, Cr_2O_3, by carbon to form chromium and
      carbon monoxide.

      .... [1]

Language lab

Use the clues below to do this crossword.

Across: 1. Colour of gas seen when copper(II) nitrate is heated

3. Solid formed when Group I nitrate is heated

4. Gas formed by heating Group I nitrate

6. Liquid formed when zinc hydroxide is heated

8. To break down a substance

Down: 2. Calcium .... is formed when calcium carbonate decomposes.

5. .... gen dioxide is formed when Group II nitrates are heated

7. Symbol for the element with the most unstable carbonate in Group II

[8]

1. The graph below shows how the percentage decomposition of four carbonates changes with time.

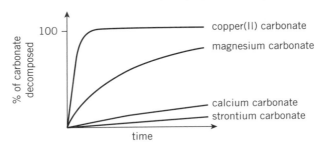

a. Put these carbonates in order of increasing rate of thermal decomposition.

.... [1]

b. How does the ease of decomposition depend on the reactivity of the metal in the carbonate?

.... [1]

2. Link the compounds **A** to **E** with the thermal decomposition products **1** to **5**.

A sodium nitrate	1 oxide + water
B copper(II) nitrate	2 nitrite + oxygen
C zinc hydroxide	3 oxide + carbon dioxide
D magnesium carbonate	4 does not decompose
E sodium hydroxide	5 oxide + nitrogen dioxide + oxygen

[3]

Complete these sentences using words from the list.

blast   blende   carbon   electrolysis   monoxide   purer   reduces   roasted   sulfide

Zinc is extracted from the ore zinc ...., which contains zinc .... In one method, the

ore is .... in air to form zinc oxide. The zinc oxide is then heated in a .... furnace

with .... The carbon burns to form carbon ...., which ....

the zinc oxide to zinc. Most zinc is now produced by the .... of zinc sulfate. This produces

.... zinc.   [9]

1. The table shows the order of some metals in the reactivity series. It also shows the position of carbon.

metal	reactivity	extracted by	energy needed to extract the metal	cost of extraction
lithium				
calcium				
cerium				
aluminium				
carbon				
zinc				
lead				
copper				
silver				
gold				

a. Draw an arrow in the second column of the table to show the reactivity of the metals
   (least reactive → most reactive).   [1]

b. Complete the third column of the table to show which metals are extracted by heating their oxides with carbon
   and which are extracted by electrolysis.   [2]

c. i. In the fourth and fifth columns draw arrows to show the amount of energy needed to extract the metal
       (less energy → more energy) and the cost of extraction (lower cost → higher cost).   [2]

   ii. Explain why there might be exceptions in the order in part **c.i.**

   .... [1]

d. Which elements in the table can be found 'native' (not combined in compounds)?

   .... [1]

2. Complete the equation for the oxidation of zinc sulfide to form zinc oxide.

   .... ZnS + ...   →   .... ZnO + ... SO_2 [2]

Search for the names of eight words related to the extraction of iron. Words go forwards or downwards only. Write the names in the space below.

B	N	H	H	A	T	E	S	C
L	S	E	C	O	K	E	X	A
A	P	M	S	X	P	Q	Q	R
S	C	A	T	I	R	O	N	B
T	G	T	E	D	R	R	S	O
X	O	I	L	E	V	F	L	N
V	E	T	L	V	B	Y	A	L
N	L	E	E	O	R	I	G	N
R	E	D	U	C	T	I	O	N

[8]

1. The diagram below shows a blast furnace for the extraction of iron.

   On the diagram draw arrows and the following letters to show:

   A → where air is blown into the furnace

   B → where the iron ore is added to the furnace

   C → where the molten iron is removed

   D → where the slag is removed

   E → where waste gases exit the furnace   [5]

2. Iron(III) oxide is reduced in the furnace by carbon monoxide.

   What are the two stages in the formation of this carbon monoxide?

   ....

   .... [2]

3. Complete these sentences about the use of limestone in the blast furnace using words from the list.

   floats   impurity   molten   oxide   silicate   silicon   slag   thermal

   Limestone undergoes .... decomposition to form calcium .... This reacts with

   .... dioxide (sand), which is an .... in the ore. The calcium ....

   (....................) formed runs down the furnace and .... on top of the .... iron.   [8]

4. Complete this equation for the reduction of iron oxide to iron.

   $Fe_2O_3(s)$ + ... CO →   ....   + ...   [2]

Language lab

Use the clues below to do this crossword.

Across:
1. .... dioxide is an impurity in iron ore
3. The large 'bucket' used in steel-making
5. Symbol for metallic element in lime
6. Lime is this type of oxide
8. The element that is blown onto the molten iron in steel-making (word is reversed)

Down:
1. The first 3 letters of an impurity in steel that has an A_r of 32
2. Phosphorus forms an .... when it combines with oxygen
3. Element present in most steels
4. A .... of oxygen reacts with the impurities in steel (word is reversed)
7. State of sulfur dioxide at r.t.p. in reverse

[10]

1. Complete the phrases **A** to **F** about steel-making on the left using phrases **1** to **6** on the right.

A Calcium oxide is added	**1** ..... is poured into a basic oxygen converter.
B A jet of oxygen	**2** ..... that floats on the iron and is removed.
C The molten iron from the blast furnace	**3** ..... are solid acidic oxides.
D Calcium silicate is a slag	**4** ..... the acidic oxides of silicon and phosphorus.
E The calcium oxide reacts with	**5** ..... because it is a basic oxide.
F The oxides of phosphorus and silicon	**6** ..... oxidises the impurities C, Si, P, and S to their oxides.

[3]

2. Complete these equations for some reactions occurring during steel-making.

   a. .... P + ... O_2 → .... P_2O_5 [1]

   b. .... + ... → $CaSiO_3$ [1]

3. Give two reasons why small amounts of chromium and manganese are added to steel after impurities have been removed.

   ....

   .... [2]

Language lab

Use the clues below to do this crossword.

Across: 1. Property of metals meaning 'can be beaten into different shapes'

3. Alloy of copper and zinc

5. Abbreviation for direct current

7. A mixture of a metal with other metals or non-metals

8. Symbol for aluminium

9. Metal used for electrical wiring

Down: 1. Type of steel used to make cars

2. Symbol for the least reactive Group I metal, which is used in aerospace alloys

3. Alloy of copper and tin

4. Iron alloy with varying amounts of carbon

6. High- .... steel is hard so is used to make hammers

[11]

1. Complete the table about the uses and properties of different metals.

metal	use	properties which makes it suitable for the use
aluminium	drinks can	1. .... [1] 2. .... [1]
copper	1. electrical wiring 2. saucepan base	1. .... [1] 2. .... [1]
mild steel (99.7% Fe, 0.3% C)	1. .... [1] 2. .... [1]	.... [1]
stainless steel (70% Fe, 20% Cr, 10% Ni)	1. .... [1] 2. .... [1]	.... [1]
tungsten steel (95% Fe, 5% W)	drill bits	1. .... [1] 2. .... [1]

Language lab

Complete these sentences about water treatment using words from the list.

bacteria   branches   filter   harmful   insoluble   plant   settle

In a water treatment ...., large objects such as plant .... are first trapped

by metal screens. Other solid particles are then left to .... to the bottom of the

tank. The water is then passed through a .... made of sand or gravel. This removes

small .... particles. Chlorine is added to the filtered water to kill

...................................., which may be .... to health.   [7]

1. The table shows the concentration in mg/dm^3 of some ions present in water from three different sources.

ion	seawater	rainwater	river water
Na^+	10 000	9	11
Ca^{2+}	900	2	1
K^+	1 000	1	4
SiO_3^{2-}	500	0.5	7
Cl^-	17 000	16	12
HCO_3^-	700	3	2
NO_3^-	trace	0.01	3

a. Which positive ion in seawater in the table is present at the lowest concentration?

.... [1]

b. What are the major differences between rainwater and river water in terms of the concentration of the ions present?

....

....

.... [3]

c. Which ion in river water is most likely to be a harmful pollutant and what is the most likely source of this ion?

.... [2]

d. Calculate the concentration of chloride ions in 200 cm³ of the river water.

Concentration = ... mg/dm^3 [1]

e. Give the chemical names of these ions:

NO_3^- .... SiO_3^{2-} .... [2]

Complete these sentences about separating gases from the air using words from the list.

distillation   first   fractional   krypton   liquid   lower   warmed

Oxygen is separated from nitrogen by .... distillation of liquid air. When liquid air is

..........................., nitrogen boils off .... because it has a .... boiling point than

oxygen. This leaves .... oxygen, which is separated from argon, ..., and

xenon by further ....   [7]

1. The table shows the percentage of some of the gases in dry air in 1985 and 2015.

gas	volume in 100 cm³ air in 1985	volume in 100 cm³ air in 2015
nitrogen	78.082	78.081
oxygen	20.950	20.949
carbon dioxide	0.034	0.0397
neon	0.0018	0.0018
helium	0.000524	0.000524
methane	0.00014	0.000179

a. Which gases have shown the largest percentage change since 1985?

.... [1]

b. Which gas present in the air at about 1% by volume is not shown in the table?

.... [1]

c. What other substance, which is liquid at r.t.p., is present in variable amounts in the air?

.... [1]

2. The apparatus shown was used to deduce the percentage of oxygen in the air.

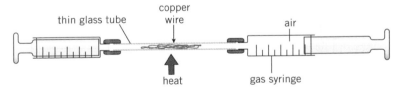

80 cm³ of air was drawn into the right-hand syringe. The air was then passed over the heated copper until there was no further decrease in volume. The final volume of air measured immediately in the right-hand syringe was 62.9 cm³.

a. What volume of oxygen had reacted? .... [1]

b. Calculate the % of oxygen in this sample of air. Show your working.

% O_2 in the air = ...%   [1]

Use the clues below to do this crossword.

Across:　2. .... burns to form SO_2

　　　　　6. The combustion is said to be ....
　　　　　 when CO is formed on burning

　　　　　7. This type of chemical reacts with limestone
　　　　　 to give off CO_2

　　　　　8. SO_2 dissolves in this to form an acid

Down:　1. Green organisms that are damaged by
　　　　　 acid rain

　　　　　3. Rain below this pH value is described as
　　　　　 acid rain

　　　　　4. Sulfur .... contributes to acid rain

　　　　　5. Element once added to petrol that can
　　　　　 cause brain damage to children

[8]

1. **a.** Complete these sentences about the sources of carbon monoxide and sulfur dioxide in the atmosphere using words from the list below. Not all the words are used.

　　　　burn　carbon　excess　fossil　gaseous　limited　oxygen　sulfur

Carbon monoxide is formed when .... compounds .... in a

.... supply of air. Sulfur dioxide is formed when .... fuels containing

.... burn in air.　　　　　　　　　　　　　　　　　　　　　　　　[5]

2. Give one harmful effect of:

　　a. An aqueous solution of nitrogen dioxide on buildings made of limestone.

　　.. [1]

　　b. Carbon monoxide on humans. .... [1]

3. Sulfur dioxide is oxidised to sulfur trioxide in the atmosphere.

　　a. How is acid rain formed from sulfur trioxide?

　　.. [1]

　　b. Describe and explain the effect of acid rain on a building made of limestone.

　　..

　　.. [3]

　　c. Describe one effect of acid rain on plants.

　　.. [1]

4. Suggest why fewer people in the world got lead poisoning in 2015 than in 1975.

　　.. [1]

Search for the names of seven words related to catalytic converters, including names of substances. Words go forwards or downwards only. Write the names in the space below.

[7]

M	O	N	O	X	I	D	E	G
X	C	C	O	P	P	I	U	A
C	A	T	A	L	Y	S	T	S
R	J	R	R	A	S	C	V	E
E	A	E	F	T	E	F	G	S
D	R	D	N	I	T	R	I	C
O	G	U	H	N	E	R	T	I
X	L	C	Y	U	R	I	T	B
E	D	E	L	M	T	I	R	F

1. The graph below shows the concentration of nitrogen dioxide in the air in the streets of a large city throughout a particular day.

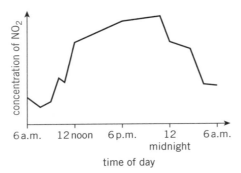

Describe and explain the shape of this graph.

....

....

.... [4]

2. Explain the function of a catalytic converter attached to a car exhaust.

....

.... [3]

3. a. State two sources of nitrogen oxides in the air.

....

.... [2]

   b. Give one harmful effect of nitrogen oxides on health.

.... [1]

4. Complete these equations for the reactions in a catalytic converter.

   a. .... NO_2 → .... + ... O_2 [2]

   b. .... NO + ... CO → .... + ...   [2]

Complete these sentences about methane using words from the list.

absorbs   atmosphere   bacterial   digestive   global   heat   vegetation   waste

Methane is a greenhouse gas formed by the .... decomposition of

.... and as a .... product in the .... system of animals. It is

present in the .... at a lower concentration than carbon dioxide but it ....

much more heat energy per mole. It traps the .... in the atmosphere, which leads to

.... warming.   [8]

1. The graphs show the concentration of carbon dioxide in the atmosphere and the estimated global mean temperature over a period of 120 years.

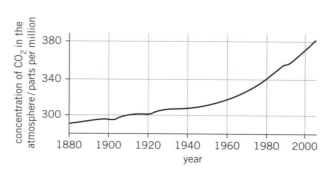 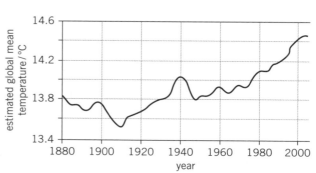

a. Carbon dioxide is a greenhouse gas. What is the meaning of the term *greenhouse gas*?

    .... [2]

b. How does the information from the graphs support the idea that carbon dioxide is a greenhouse gas?

    ....

    .... [2]

c. What evidence is there from these graphs that global warming might not always be related to the concentration of carbon dioxide in the atmosphere?

    ....

    .... [2]

2. The absorption of energy by greenhouse gases leads to global warming.
    Give three effects of global warming.

    ....

    ....

    .... [3]

Use the clues below to do this crossword.

Across: 2. Photo .... produces most of the oxygen in the atmosphere

3. The release of energy from glucose in living things

6. A fossil .... when burnt releases CO_2

7. The number of C atoms in one molecule of glucose

8. A product of photosynthesis

Down: 1. Breakdown of vegetation by bacteria

2. Sea creatures such as mussels have .... made out of calcium carbonate

4. Large seas that absorb huge amounts of carbon dioxide

5. A solid fossil fuel

[9]

1. The diagram shows the carbon cycle. The numbers are the relative amounts of carbon transferred per year.

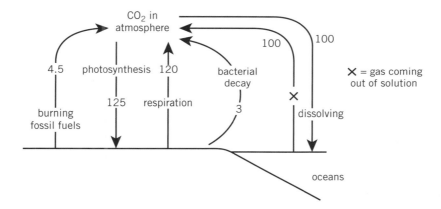

a. What are the two main processes releasing carbon dioxide into the atmosphere?

.... [1]

b. What are the two main processes removing carbon dioxide from the atmosphere?

.... [1]

c. Comment on the balance between these processes.

....

.... [2]

d. Describe two other factors that might upset the balance of the carbon cycle.

....

.... [2]

Language lab

Complete these sentences about rusting using words from the list.

corrodes　electrons　ions　iron　more　rusting　sacrificial　solution

Blocks of zinc can be placed on the hull of a ship to stop it .... Zinc is ....

reactive than .... so it loses .... and forms .... more easily than iron. The

zinc ions go into .... and so the zinc .... instead of the iron. This is called

.... protection.　　　　　　　　　　　　　　　　　　　　　　　　　[8]

1. The graph shows the rate of corrosion of iron at different pH values in aerated water.

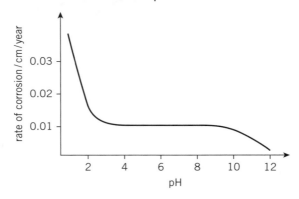

    a. Describe how the corrosion of iron varies with pH.

    ....
    .... [3]

    b. At neutral pH values H^+ ions are used up leaving OH^- ions in solution. These first react with iron(II) ions to form iron(II) hydroxide ('green rust').
        Complete the ionic equation for this reaction. Include state symbols.

        Fe^{2+}(aq)　+　............................　→　............................　　　　[3]

    c. At more alkaline pH values 'green rust' is converted to 'red rust' (hydrated iron(III) oxide).

        i.　Give the name of the oxidising agent in this reaction. .... [1]

        ii.　'Green rust' is converted to 'red rust' more rapidly at more alkaline pH values. What information from the graph suggests that 'red rust' protects the iron from corrosion better than 'green rust'?

        .... [1]

2. Suggest reasons for the following:

    a. An iron object in the desert rusts very slowly.

    .... [1]

    b. Painting an iron object stops it from rusting.

    ....
    .... [2]

Language lab

Complete these sentences about fertilisers using words from the list.

fertilisers   nitrates   nutrients   phosphates   phosphorus   proteins   salts

For healthy growth, crop plants need three major elements: nitrogen, ..., and

potassium. Plants take up these elements in the form of nitrates, ..., and

potassium .... The .... are needed to make .... for growth.

Farmers add .... to the soil to add back the .... that plants

have absorbed for growth.   [7]

1. A flow chart for making fertilisers is shown below.

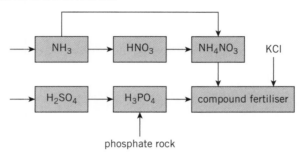

a. Name the compounds in the diagram:

NH_3 .... HNO_3 ....

NH_4NO_3 .... H_2SO_4 ....

H_3PO_4 ....   KCl ....   [6]

b. Write a word equation for the reaction between NH_3 and HNO_3.

.... [1]

c. Name a suitable acid and base for making these fertilisers.

  i. Ammonium sulfate .... [2]

  ii. Potassium chloride .... [2]

  iii. Sodium phosphate .... [2]

2. a. Farmers add lime to their fields. Suggest why they do this.

.... [1]

b. Lime reacts with water in the soil to form calcium hydroxide, $Ca(OH)_2$. This can react with ammonium sulfate fertiliser, $(NH_4)_2SO_4$, in the soil. Write a chemical equation for this reaction.

.... [2]

Language lab

Search for the names of seven words related to the manufacture of ammonia. Words go forwards or downwards only. Write the names in the space below.

C	A	T	A	L	Y	S	T	H
X	C	P	T	O	S	D	F	Y
Q	W	E	I	R	T	Y	U	D
N	I	T	R	O	G	E	N	R
A	H	D	O	F	G	H	J	O
R	A	W	N	C	V	B	M	G
F	B	M	E	T	H	A	N	E
V	E	E	D	C	T	Y	U	N
P	R	E	S	S	U	R	E	C

[7]

1. Give two ways by which hydrogen is produced to make ammonia.

.... [2]

2. Complete the equation for the synthesis of ammonia.

$N_2(g) +$ .... (g) $\rightleftharpoons$ .... (g)   [2]

3. The graph below shows the effect of temperature and pressure on the percentage yield of ammonia.

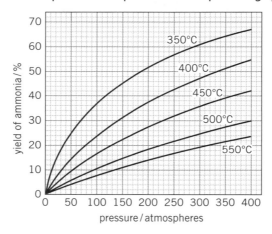

a. Describe the effect of pressure on the percentage yield.

.... [1]

b. How do the data in the graph show that the reaction is exothermic?

.... [1]

c. What is the percentage yield of ammonia at 200 atmospheres and 350 °C?

.... [1]

d. State one advantage and one disadvantage of using a low temperature in the reaction.

....

.... [2]

Use the clues below to do this crossword.

Across: 1. One of the products when H_2SO_4 reacts with zinc oxide

3. The .... of 0.1 mol/dm³ H_2SO_4 is between 1 and 2

6. Formula for zinc sulfide

8. A major use of sulfuric acid is to make this mixture, which contains N, P, and K

10. Ion with the formula SO_3^{2-}

11. H_2SO_4 is used to make the colours for this 'liquid' we brush on walls and doors

Down: 2. Number of moles of H_2SO_4 in 980 g of H_2SO_4

4. Gas released when H_2SO_4 reacts with zinc

5. SO_2 is used to .... wood pulp

7. Acid + base → .... + water

9. The colour is taken out of wood .... by using SO_2

[11]

1. Petroleum contains unwanted sulfur compounds.

a. Why is it important that these sulfur compounds do not remain in fuels obtained from the fractional distillation of petroleum?

....

.... [2]

b. The sulfur compounds are converted to hydrogen sulfide by reduction with hydrogen using a catalyst. What is the purpose of the catalyst?

.... [1]

c. The hydrogen sulfide is then separated from other gases by reacting it with an organic solvent. Suggest why the solvent is able to separate hydrogen sulfide from the other gases.

.... [1]

2. Write word equations and balanced chemical equations for these reactions:

a. The reaction of sulfuric acid with potassium hydroxide.

....

.... [3]

b. The reaction of sulfuric acid with magnesium.

....

.... [2]

Language lab

Search for the names of seven words related to the manufacture of sulfuric acid. Words go forwards or downwards only. Write the names in the space below.

A	B	C	O	N	T	A	C	T	Z
R	T	Y	C	U	R	S	H	R	R
V	A	N	A	D	I	U	M	X	R
O	Q	V	T	T	O	A	S	D	F
L	W	B	A	O	X	Y	G	E	N
E	E	N	L	Y	I	A	S	D	F
U	R	M	Y	H	D	D	F	G	H
M	T	S	S	J	E	H	J	K	G
E	X	O	T	H	E	R	M	I	C

[7]

1. In the Contact process, sulfur dioxide is converted to sulfur trioxide in the presence of vanadium(V) oxide.

$$2SO_2(g) + O_2(g) \rightleftharpoons 2SO_3(g)$$

a. What is the purpose of the vanadium(V) oxide?

.... [1]

b. The graph below shows the percentage conversion of SO_2 to SO_3 at different temperatures. The pressure was just above atmospheric pressure.

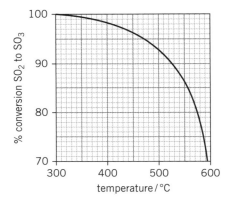

i. Describe in detail the effect of temperature on the percentage conversion of SO_2 to SO_3.

....

.... [2]

ii. Deduce the percentage conversion of SO_2 to SO_3 at 500 °C.

.... [1]

iii. How do the data in the graph show that the reaction is exothermic?

.... [1]

c. Predict the effect of increasing the pressure on this reaction. Explain your answer.

....

.... [2]

Complete these sentences about flue gas desulfurisation using words from the list.

carbonate   combustion   fuels   neutralise   sulfite   sulfur   waste

Flue gas desulfurisation is the process of removing .... dioxide from the gases formed

during the .... of fossil .... in power stations. The .... gases are

passed through moist calcium .... or calcium oxide. These compounds

.... the acidic sulfur dioxide. Solid calcium .... is formed.   [7]

1. The diagram shows an old-fashioned kiln for making calcium oxide from calcium carbonate.

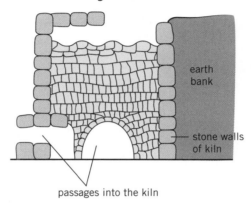

earth
bank

stone walls
of kiln

passages into the kiln

a. Put the following labels on the diagram:

   T to show where the limestone is tipped into the kiln.

   F to show where the fuel (coal) is being burnt.   [2]

b. How does air get into the kiln and why is it needed?

   ....

   .... [3]

c. The walls of the kiln are made of granite. Granite is mainly a mixture of silicon dioxide and aluminium oxide.
   Why are the walls made of granite and not of limestone?

   ....

   .... [2]

d. The equation shows the thermal decomposition of calcium carbonate.

$$CaCO_3(s) \rightleftharpoons CaO(s) + CO_2(g)$$

   i. What is meant by the term *thermal decomposition*?

      .... [1]

   ii. Explain why the backward reaction is unlikely to occur in the lime kiln.

      ....

      .... [3]

Link the names **A** to **D** on the left with the descriptions **1** to **4** on the right.

A Alcohols	**1** A group of organic compounds having a −COOH group
B Alkanes	**2** Hydrocarbons having one or more C=C bonds
C Alkenes	**3** A group of organic compounds having an −OH group
D Carboxylic acids	**4** A group of hydrocarbons with only single bonds [2]

1. a. What is meant by the term homologous series?

    ....

    .... [2]

    b. To which homologous series do these compounds belong?

    Propene .... Butanol ....

    Hexane .... Propanoic acid .... [4]

2. Write the molecular formula for each of these compounds.

    Ethane .... Ethanol ....

    Ethanoic acid .... Ethene .... [4]

3. Write the full structural formulae for each of these compounds, showing all atoms and bonds.

a. Propane, $CH_3CH_2CH_3$	b. Propanol, $CH_3CH_2CH_2OH$
c. But-1-ene, $CH_3CH_2CH=CH_2$	d. Ethanoic acid, CH_3COOH

[4]

Use the clues below to do this crossword.

Across: 3. Alkane with three carbon atoms

4. The number of carbon atoms in hexane

5. The homologous series with the general formula C_nH_{2n+2}

7. .... mers have the same molecular formula but different structural formulae

8. C_2H_4 and C_3H_6 are examples of these

Down: 1. Number of carbon atoms in butane

2. One of the elements in a hydrocarbon

4. Ethene and propene are both in the same homologous ....

6. C_2H_5- is an example of one of these groups

9. Meth- indicates one carbon atom in an organic compound. What prefix indicates two carbon atoms?

[10]

1. What is the meaning of the term *hydrocarbon*?

.... [1]

2. Name the alkanes with unbranched chains of:

   a. Five carbon atoms .... [1]

   b. Four carbon atoms .... [1]

   c. Six carbon atoms .... [1]

3. Name the alkenes with unbranched chains of:

   a. Three carbon atoms .... [1]

   b. Two carbon atoms .... [1]

   c. Five carbon atoms .... [1]

4. Draw two isomers of the hydrocarbon with the formula C_5H_{12}. [2]

Use the clues below to do this crossword.

Across: 4. A liquid which vaporises easily is described as ....

5. A product of the combustion of ethane

6. Symbol for one of the elements in a compound used to test for water

7. Fuel used for lorries and cars

9. The physical state of coke at r.t.p.

Down: 1. Black solid fossil fuel

2. The main compound in natural gas

3. Another name for gasoline

4. A liquid which is thick and sticky is described as ....

8. Petroleum is separated into fractions in an oil refin ....

[10]

1. Petrol can be made from coal using the following route:

| A coal | → | B decomposition using heat and hydrogen | → | C refining based on boiling point | → | D petrol |

a. Which stage, **A**, **B**, **C**, or **D**, involves distillation? .... [1]

b. Which stage involves reduction? .... [1]

2. The table shows some properties of different petroleum fractions.

fraction	boiling point range / °C	size of molecules	volatility	ease of flow	ease of burning
1	up to 100				
2	100–150				
3	150–200				
4	200–300				

a. In the third column draw an arrow to show how the size of the molecules varies with the boiling point range (low → high). [1]

b. i. Some compounds are volatile. What is the meaning of the term volatile?

.... [1]

ii. In the fourth column draw an arrow to show how the volatility of the compounds varies with the boiling point range (low → high). [1]

c. In the fifth column draw an arrow to show how the viscosity ('syrupiness') of the compounds varies with the boiling point range (flows easily → flows less easily). [1]

d. In the sixth column draw an arrow to show how the ease of burning of the compounds varies with the boiling point range (difficult → easy). [1]

Complete these sentences about the distillation of petroleum using words from the list.

boiling   bottom   condense   further   higher   lower   temperatures   top

There is a range of .... in the distillation column, hot at the .... and cooler

at the .... Hydrocarbons with .... boiling points move .... up the

column and .... when the temperature in the column falls just below the

.... point of the hydrocarbons. Hydrocarbons with .... boiling points

condense lower down the column.   [8]

1. The diagram shows a simple experiment to separate petroleum fractions.

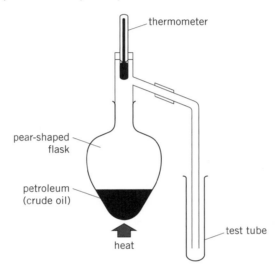

Describe how you could use this apparatus to separate the hydrocarbon fractions.

....

....

....

.... [4]

2. The formulae of some compounds found in petroleum are shown below.

A	B	C	D

a. Which one of these compounds is a branched-chain compound? .... [1]

b. Which one of these compounds is a ring compound? .... [1]

Complete these sentences about the reaction of alkanes with chlorine.

| chlorine | combustion | excess | green | hydrogen |

| photochemical | substitution | sunlight | unreactive |

Alkanes are generally .... except for the reaction with chlorine in the presence of

light (a .... reaction) and .... When .... alkane is mixed

with chlorine in a sealed tube and exposed to ...., the .... colour of the

chlorine disappears. A chlorine atom replaces a .... atom in the alkane. This type of

reaction is called a .... reaction. If excess .... is present, more than one

hydrogen atom is replaced by chlorine. [9]

1. Complete these sentences about alkanes.

    a. Alkanes are .... because they only have hydrogen and carbon atoms in their structure. [1]

    b. All the bonds in alkanes are .... bonds. [2]

2. How do the boiling points of the alkanes change with relative molecular mass?

    .... [1]

3. Complete these equations for the typical reactions of alkanes.

    a. C_5H_{12} + ... O_2 → .... CO_2 + ... H_2O [2]

    b. .... C_4H_{10} + ... → .... CO + ... H_2O [2]

    c. CH_4 + Cl_2 → .... + HCl [1]

    d. C_2H_6 + ... Cl_2 → $C_2H_4Cl_2$ + ... [2]

4. Which two words describe the reaction in **3.c**? Put rings around the correct answers.

    addition   cracking   neutralisation   photochemical   polymerisation   substitution [2]

5. Draw two isomers of the hydrocarbon with the formula C_5H_{12} showing all atoms and all bonds. [2]

Language lab

Use the clues below to do this crossword.

Across: 1. Breakdown of alkanes into smaller alkanes and alkenes by heat

3. Alkene formed by decomposing ethane

5. A gas formed by cracking that is not a hydrocarbon

7. Cracking is a type of .... decomposition

Down: 1. This speeds up cracking

2. $C_{16}H_{34} \rightarrow A + C_7H_{14}$
How many C atoms does A have?

4. A condition needed for cracking

6. Aluminium .... is a catalyst used in cracking

[8]

1. The bar chart shows the supply and demand for different petroleum fractions.

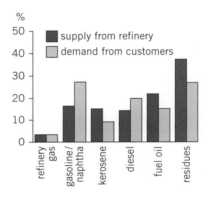

a. Which of the fractions shown has molecules with the longest chains?

.... [1]

b. Name and give the formula of one molecule present in refinery gas.

.... [2]

c. i. For which fractions is the demand much greater than the supply?

.... [1]

ii. For which fractions is the supply much greater than the demand?

.... [1]

3. Complete these equations for cracking.

a. $C_{10}H_{22} \rightarrow C_4H_{10} + \ldots\ldots\ldots$ [1]

b. $C_{14}H_{30} \rightarrow C_3H_8 + C_4H_8 + \ldots\ldots\ldots$ [1]

Complete these sentences about saturated and unsaturated compounds using words from the list.

  aqueous **decolourised** **orange** **remains** **saturated** **unsaturated**

We can tell the difference between an unsaturated and a .... compound by adding .... bromine to a sample of the compound. Aqueous bromine is .... in colour. If the aqueous bromine is ...., the compound is .... If the bromine water .... orange, the compound is saturated. [6]

1. **a.** The table gives some information about the alkenes. Complete the table in the spaces provided.

name of alkene	molecular formula	boiling point / °C
ethene	....	−102
....	C_3H_6	....
....	....	−7
pentene	....	30
hexene	C_6H_{12}	....

[7]

 b. Which of the alkenes in the table are likely to be liquids at r.t.p.?

.... [1]

2. The structure of compound **T** is shown below.

$$H_2N-CH=CH-CH_2-O-H$$

On the structure of **T** above draw a ring around the functional group which makes this compound unsaturated. [1]

3. Complete the boxes in the diagram below to show the structure of **A** and formula of the additional reagent **B** including the state symbol. [2]

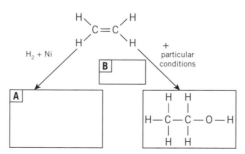

Language lab

Use the clues below to do this crossword.

Across: 2. A product formed when alcohols burn

6. Alcohol with four carbon atoms

7. Ethanol has .... carbon atoms

8. Abbreviation for oxidation

9. Name of C_2H_5OH

Down: 1. The C_2H_5- group

3. Heating with the condenser in the upright position

4. Sugar used in fermentation to produce ethanol

5. Organism used in fermentation to produce ethanol

8. Liquid extracted from petroleum or from plants

[10]

1. The diagram shows the apparatus used to convert ethanol to ethanoic acid.

a. On the diagram, draw an arrow to show where heat is applied. [1]

b. What type of reagent is the potassium manganate(VII)?

.... [1]

water out

water in

ethanol + potassium manganate(VII) + acid

c. What colour change would you expect to see if the potassium manganate(VII) was not in excess?

from ....

to .... [2]

d. Why is the condenser in the upright position?

....

....

.... [2]

2. a. Write a chemical equation for the complete combustion of ethanol.

.... [2]

b. Complete the equation for the oxidation of ethanol to ethanoic acid.

.... + 2[O] → .... + ... [2]

Language lab

Complete these sentences about esterification using words from the list.

alcohol   catalyst   flavourings   insoluble   neutralise   separating   sulfuric

Esters are used in .... and perfumes. Esters are made by heating an ....

with a carboxylic acid. A few drops of .... acid are added to act as a

.... After the reaction is complete, excess sodium carbonate is added to

.... any excess acid. Esters are .... in water, so they can be

separated from the rest of the reaction mixture using a .... funnel.   [7]

1. Ethanoic acid dissolves in water.

$$CH_3COOH + H_2O \rightleftharpoons CH_3COO^- + H_3O^+$$

   a. How does this equation show that ethanoic acid is a weak acid?

   ....

   .... [2]

   b. Explain why water is acting as a base in this reaction.

   .... [2]

2. Complete these equations for some reactions of ethanoic acid.

   a. .... CH_3COOH + ... Na → .... + ...   [3]

   b. .... CH_3COOH + Mg → .... + ...   [3]

   c. CH_3COOH + NaOH → .... + ...   [2]

   d. .... + ... → CH_3COOCH_3 + ...   [3]

3. Draw the structure of these esters showing all atoms and all bonds.

a. ethyl butanoate	b. propyl methanoate
[1]	[1]

4. Name these esters:

   a. $HCOOC_4H_9$ .... b. $C_2H_5COOC_2H_5$ .... [2]

Complete these sentences about polymers using words from the list.

bonds   ethene   join   molecules   monomers   polymerisation

A polymer is a substance that has very large ...., formed when lots of small molecules

called .... join together. This process is called .... When poly(ethene)

is formed, one of the C=C .... of .... is broken and the monomers

.... together in a chain.   [6]

1. In the box below draw a section of the polymer chain formed by the addition of four units of ethene monomers.

[2]

2. Many plastics are non-biodegradable.

    What is meant by the term *non-biodegradable*?

    ....

    .... [2]

3. We can get rid of waste plastics by recycling, putting them in landfill sites, or by burning them.

    a. Give one advantage of recycling plastics.

        .... [1]

    b. Give one advantage of putting plastics in landfill sites.

        .... [1]

    c. Give one advantage of burning plastics.

        .... [1]

Complete these sentences about addition polymers using words from the list.

addition   bonds   combine   formed   monomers   no   two

When .... containing C=C double .... are polymerised, no other molecule

is .... apart from the polymer. We call this type of polymerisation ....

polymerisation. An addition reaction is a reaction in which .... or more molecules

.... and .... other molecule is formed.   [7]

1. The structure of a monomer is shown below.

$$H_3C \diagdown \atop H \diagup C = C \diagup^H \diagdown_F$$

Draw a section of the polymer chain formed from this monomer. Show three repeat units.

[3]

2. Draw the structure of the polymer formed from but-2-ene, $CH_3-CH=CH-CH_3$, as one repeat unit of this polymer with brackets and *n*.

[3]

3. Draw the monomers of polymers **A** and **B**.

A

$$-\underset{\underset{H}{|}}{\overset{\overset{C_6H_5}{|}}{C}}-\underset{\underset{H}{|}}{\overset{\overset{H}{|}}{C}}-\underset{\underset{H}{|}}{\overset{\overset{C_6H_5}{|}}{C}}-\underset{\underset{H}{|}}{\overset{\overset{H}{|}}{C}}-\underset{\underset{H}{|}}{\overset{\overset{C_6H_5}{|}}{C}}-\underset{\underset{H}{|}}{\overset{\overset{H}{|}}{C}}-\underset{\underset{H}{|}}{\overset{\overset{C_6H_5}{|}}{C}}-$$

B

$$-\underset{\underset{H}{|}}{\overset{\overset{H}{|}}{C}}-\underset{\underset{H}{|}}{\overset{\overset{CN}{|}}{C}}-\underset{\underset{H}{|}}{\overset{\overset{H}{|}}{C}}-\underset{\underset{H}{|}}{\overset{\overset{CN}{|}}{C}}-\underset{\underset{H}{|}}{\overset{\overset{H}{|}}{C}}-$$

[4]

Language lab

Complete these sentences about condensation polymerisation using words from the list.

> chloride   eliminated   functional   small   water

In condensation polymerisation, molecules with different .... groups

react together. A .... molecule such as .... or hydrogen ....

is ....   [5]

1. The diagram shows two polymers, **A** and **B**.

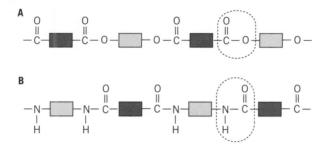

   a. On each of the diagrams above put brackets to show one repeat unit.   [2]

   b. Give the name of the linking group in

   i. Polymer **A** ....   ii. Polymer **B** ....   [2]

2. The structure of poly(lactic acid) is shown below.

$$-O-\underset{\underset{H}{|}}{\overset{\overset{CH_3}{|}}{C}}-\overset{\overset{O}{||}}{C}-O-\underset{\underset{H}{|}}{\overset{\overset{CH_3}{|}}{C}}-\overset{\overset{O}{||}}{C}-O-\underset{\underset{H}{|}}{\overset{\overset{CH_3}{|}}{C}}-\overset{\overset{O}{||}}{C}-$$

   a. Give the names of the two functional groups that react to form this polymer.

   .... and ....   [2]

   b. Draw the structure of the single monomer that is used to form this polymer.

                                                                                                              [2]

Use the clues below to do this crossword.

Across: 2. Boiling a reaction mixture in apparatus that cools the vapour and returns it to the flask

4. A class of food material containing carbon, hydrogen, nitrogen, and oxygen

6. The linkage group in proteins

9. Amino acids .... together to form proteins

Down: 1. A carbohydrate with the formula $C_6H_{12}O_6$

3. .... acids are formed when proteins are hydrolysed

5. A catalyst present in the cells of the body

6. A substance used to hydrolyse proteins

7. When amino acids join, water is eli_ _ _ at _ d

8. The symbols of the atoms in the group present in both esters and amides

[10]

1. The diagram below shows part of a protein.

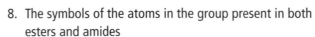

a. On the diagram above, put brackets to show one repeat unit. [1]

b. Give the name of the linking group. .... [1]

2. The structure of part of poly(glycine) is shown below.

$$-N-C-CH_2-N-C-CH_2-N-C-$$

Deduce the structure of the monomer that is used to form this polymer.

[2]

3. Proteins are hydrolysed to amino acids. What is the meaning of the term hydrolysis?

.... [2]

Complete these sentences about the chromatography of carbohydrates using words from the list.

chromatography   coloured   front   locating   mixture   paper   position

The different carbohydrates in a .... can be identified using .... chromatography.

After carrying out chromatography, the position of the solvent .... is marked. The

.... paper is then sprayed with a .... agent. This makes the

.... of each carbohydrate visible as a .... spot.   [7]

1. The diagram below shows part of a starch molecule.

   a. How many repeat units are shown? .... [1]

   b. What type of polymerisation has taken place when starch is formed from glucose?

       .... [1]

   c. Draw a block diagram to show the structure of the monomer.

                                                                                                    [1]

2. The block diagram of fructose is

   Draw a diagram to show part of the polymer of fructose. Show four repeat units.   [3]

3. The diagram shows the apparatus used for hydrolysing starch.

   a. Put arrows on the diagram to show where water enters
       the condenser and where it goes out.   [1]

   b. Why is the condenser in the vertical position?

       ....

       ....

       .... [2]

Language lab

Search for the names of six words about the production of ethanol by fermentation, including names of reactants and products. Words go forwards or downwards only. Write the names in the space below.

X	E	T	H	A	N	O	L	A
G	L	U	C	O	S	E	Y	S
C	E	Q	Y	W	E	R	T	D
A	N	A	E	R	O	B	I	C
R	Z	Z	A	V	X	C	V	F
U	Y	W	S	A	T	N	B	A
O	M	E	T	F	Y	U	I	Y
N	E	A	S	D	F	G	H	J
L	S	R	D	I	S	T	I	L

[6]

1. a. Ethanol can be manufactured by fermentation or by hydration of ethene.

Complete the table about these reactions.

	fermentation	hydration of ethene
reagents needed		
temperature / °C		
pressure		
catalyst		

[8]

b. Give two disadvantages of producing ethanol by fermentation.

....

.... [2]

c. Give two advantages of producing ethanol by fermentation.

....

.... [2]

d. Give two advantages of producing ethanol by hydration of ethene.

....

.... [2]

Non-scientific words commonly used in chemistry questions can be just as important as the scientific words that you have to know.

1. Here is list of words that are commonly used in chemistry questions. Only a few of these are chemical terms. Link each of the words **A** to **H** with the words or phrases **1** to **8**.

A	Arrangement	**1**	Typical of
B	Decompose	**2**	A coloured substance
C	Characteristic of	**3**	A particular quality about something
D	Produces	**4**	Carries on
E	Converted	**5**	How things are positioned
F	Pigment	**6**	Makes
G	Property	**7**	Changed to
H	Proceeds	**8**	Break down

[4]

2. Link each of the words **I** to **P** with the words or phrases **9** to **16**.

I	Investigate	**9**	How close things are to each other
J	Distinguish	**10**	Part of something
K	Proximity	**11**	Find out about
L	Appearance	**12**	Opposite, often negative or harmful
M	Component	**13**	What something looks like
N	Combination	**14**	Shows
O	Adverse	**15**	The joining of two or more things
P	Represents	**16**	Tell the difference between

[4]

3. Link each of the words **Q** to **W** with the words or phrases **17** to **23**.

Q	Factor	**17**	Shown
R	Structure	**18**	Temperature and pressure are examples of these
S	Demonstrated	**19**	Compared with each other
T	Conditions	**20**	Sort
U	Relative	**21**	Place
V	Type	**22**	One of the things which affects a result
W	Position	**23**	The way in which something is put together

[4]

Some definitions in chemistry involve two or more different points. This exercise will help you remember how to write these definitions.

1. a. Complete this definition of the term *hydrocarbons* using words from the list.

    carbon   compounds   only

    Hydrocarbons are .... containing .... hydrogen and

    .... atoms.   [1]

    b. Which word do you think is often left out by students when defining a hydrocarbon?

    .... [1]

2. a. Complete this definition of the term *compound* using words from the list.

    bonded   different   substance   two

    A compound is a .... that contains .... or more .... atoms

    .... together.   [2]

    b. Which two words do you think are often left out by students when defining a compound?

    .... [2]

3. a. Complete this definition of the term *relative atomic mass*.

    atoms   average   element   scale   twelve

    Relative atomic mass is the .... mass of naturally occurring .... of an

    .... on a .... on which the ^{12}C atom has a mass of exactly

    .... units.   [3]

    b. Check the answer to see if you are correct, then write out the definition again and underline the five most important words in the definition. They are not all the same as the gaps!

    ....

    ....

    .... [6]

4. Complete this definition of the term *mole* using words from the list.

    atoms   amount   grams   isotope   molecules   particles

    A mole is the .... of substance that has the same number of ....

    (atoms, ions, ..., or electrons) as there are .... in exactly twelve

    .... of the carbon-12 ....   [3]

Some words are essential when describing chemical terms and writing extended answers (answers requiring writing in phrases or sentences). This exercise will help you remember these words.

In each of these sentences, cross out the incorrect word and/or write in the missing words.

1. Activation energy is the <u>maximum / minimum</u> energy that colliding particles must have for a reaction to

    take place.   [1]

2. An element is a substance made up of only one .... of atom.   [1]

3. Isotopes are <u>atoms / molecules</u> of an <u>compound / element</u> with the same number of protons but different

    numbers of neutrons.   [2]

4. Electrolysis is the .... of an ionic compound, when molten or in aqueous solution, by the

    passage of ....   [2]

5. An oxidising agent oxidises another substance when that substance <u>gains / loses</u> electrons.   [1]

6. The conduction of electricity in an electrolyte is due to the movement of <u>electrons / ions</u>.   [1]

7. When hydrochloric acid reacts with magnesium, increasing the concentration of the acid ....

    the rate of the reaction because there are more particles of acid in a given <u>area / volume</u>. So the

    <u>frequency / number</u> of collisions increases.   [3]

8. Sodium chloride conducts electricity when molten because the <u>atoms / ions</u> are free to ....   [2]

9. When sodium reacts with chlorine, each sodium <u>atom / ion</u> loses an electron. Each chlorine <u>atom / molecule</u>

    gains one electron.   [2]

10. The <u>atoms / components</u> of a mixture can be separated by <u>chemical / physical</u> means.   [2]

11. Unsaturated compounds contain carbon–carbon <u>double / single</u> bonds. When added to aqueous bromine

    the colour of the aqueous bromine changes from <u>orange / red</u> to <u>white / colourless</u>.   [3]

1. Link each of the words **A** to **F** about the properties of some metals to the meanings **1** to **6**.

A Dense	**1** Can be beaten into shape	
B Ductile	**2** Not easily changed by a force	
C Lustrous	**3** Can be drawn into wires	
D Malleable	**4** Rings when hit	
E Sonorous	**5** Has high mass to volume ratio	
F Strong	**6** Shiny	[3]

2. Link each of the terms **A** to **G** about environmental chemistry to the phrases **1** to **7**.

A Acid rain	**1** Can be broken down by organisms	
B Adverse effect	**2** A gas which warms the atmosphere	
C Biodegradable	**3** Burning in limited supply of air	
D Climate change	**4** Thick mist caused by NO_2 and hydrocarbons	
E Greenhouse gas	**5** An example of this is desertification	
F Incomplete combustion	**6** Harmful action	
G Photochemical smog	**7** Precipitation which has pH 5 or less	[4]

3. Link each of the chemical terms **A** to **F** to the meanings **1** to **6**.

A Decomposition	**1** Reaction of acid with a base to form a salt and water	
B Displacement	**2** Gain of electrons	
C Distillation	**3** Reaction which depends on light	
D Neutralisation	**4** One atom or group replacing another	
E Photochemical	**5** Separation due to different boiling points	
F Reduction	**6** Breakdown of a substance	[3]

1. Link the beginnings of these sentences **A** to **G** with the correct endings **1** to **7**.

A We can use the titration method to	**1** ..... separate a mixture of liquids with different boiling points.
B We use upward displacement of air to	**2** ..... make an insoluble salt from two soluble compounds.
C Fractional distillation is used to	**3** ..... collect gases which are insoluble in water.
D We can use downward displacement of water to	**4** ..... deliver a given volume of liquid accurately.
E We can use the precipitation method to	**5** ..... make a soluble salt from an alkali and an acid.
F A burette is used to	**6** ..... separate a solid from a solution.
G Filtration is used to	**7** ..... collect gases which are denser than air.

[4]

2. Join up these fragments to form a sentence describing why diamond has a high melting point.

    ..... between all the carbon atoms.   Diamond has a high melting point .....

    ..... covalent bonds that exist .... to break the strong .....

    ..... because it takes a lot of energy .....

    ....

    .... [2]

3. Join up these fragments to form two sentences describing how the rate of reaction changes with temperature.

    ..... the faster the particles move .... reactant particles being successful.

    ..... equal to or greater than the activation energy .... also increases, so there is .....

    The higher the temperature, ...   The number of particles having energy .....

    ..... because they have more energy and collide with .... a greater frequency.

    ..... more chance of collisions between .....

    The higher the temperature, ...

    ....

    ....

    .... [3]

An important language skill is the ability to read a passage of unfamiliar material and answer questions about the content of the passage.

1. Read this passage about photochemical smog and answer the questions that follow.

In the high temperature and pressure inside a car engine, nitrogen combines with oxygen to form harmful oxides of nitrogen and carbon monoxide. Neither of these compounds is completely removed by a catalytic converter, which converts nitrogen oxides to nitrogen and carbon monoxide to carbon dioxide. Nitrogen and carbon dioxide are harmless to health. The exhausts from car engines also contain unburnt hydrocarbons. The hydrocarbons and nitrogen oxides combine with ozone, O_3, from the atmosphere in the presence of sunlight to form photochemical smog. This is made worse in cities where a layer of warm air traps a layer of cooler air beneath it. In the presence of ultraviolet radiation, ozone reacts with nitrogen oxides in a cycle to produce oxygen and more nitrogen oxides. Hydrocarbons and carbon monoxide disrupt this cycle, and aldehydes (such as CH_3CHO and C_6H_5CHO), peroxides (such as H_2O_2), and organic nitrates are formed. These compounds can irritate the eyes, cause breathing difficulties, and make asthma sufferers very ill.

a. What is the function of a catalytic converter?

....

.... [2]

b. What is the meaning of the term *photochemical*?

.... [1]

c. Give the formula for the aldehyde functional group.

.... [1]

d. Apart from the chemicals present, what factor makes photochemical smog worse?

.... [1]

e. Give two adverse effects of photochemical smog.

....

.... [2]

f. What conditions are needed for nitrogen oxides to form in a car engine?

.... [2]

g. How many atoms are there in one molecule of ozone?

.... [1]

2. Read this question, which requires an extended answer. Underline the five most important words or phrases that you need to consider when answering this question.

When water boils to form steam there is a change of state. Explain what happens when this change of state occurs in terms of the motion and energy of the particles. [5]

At the beginning of a question is a part called the 'stem'. The stem tells you what the question is about.

- The stem may include diagrams or tables.
- Look for key words in the stem that tell you about the question.
- You might find it useful to underline key words.
- Make sure that you read every word. Sometimes it is the non-scientific words that cause trouble because when we are reading quickly we sometimes miss these out.
- Remember that you may need to refer back to the stem of the question when you answer different parts of the question.

The diagram below shows you a question stem and what to look for.

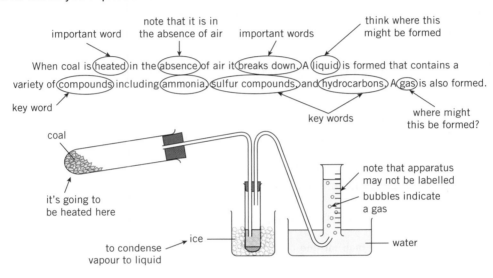

What can you find out from the stem and diagram without any questions being asked?

- It's a thermal decomposition reaction because of the words 'heat' and 'breaks down' (decomposition means breakdown).
- A gas is collected in the measuring cylinder (you need to know the names of pieces of apparatus).
- The gas is insoluble in water (otherwise it would dissolve in the water and not collect at the top).
- It's the coal that is heated. So if you are asked to draw an arrow to show where the heat is applied it should be under the coal, not anywhere else.

When reading each part of a question:

- Read each word carefully, including the non-scientific words. Don't rush through the passage.
- Look for key words, including command words. A list of the most important command words is given in units 22.1 and 22.2.
- Note the number of examples that you have to give.

1. Underline the most important words in each of these questions. The number of words to underline is given after each example.

   a. Give two examples of radioactive isotopes used in industry. (Underline 3 words.)   [3]

   b. Describe how you could separate the different coloured compounds present in a mixture of dyes.
      (Underline 3 words.)   [3]

   c. Describe what you would observe when dilute hydrochloric acid is added to zinc.
      (Underline 4 words.)   [4]

   d. Describe and explain condensation in terms of the energy and movement of particles.
      (Underline 6 words.)   [6]

2. Read these questions and the **incorrect** answers given.
   In each case describe and explain what mistake has been made in reading the question.

   a. Question: Give two properties shown by all metals.

      Incorrect answer: high density, high melting point

      Mistake: ....

      .... [2]

   b. Question: What observation would you make when nitric acid is added to zinc?

      Incorrect answer: hydrogen gas is given off

      Mistake: ....

      .... [2]

   c. Question: Give an industrial use of radioactive isotopes.

      Incorrect answer: treating cancers

      Mistake: ....

      .... [2]

   d. Question: What feature of the ethene molecule shows that it is an unsaturated hydrocarbon?

      Incorrect answer: It decolourises bromine water

      Mistake: ....

      .... [2]

Command words tell us what sort of thing we need to write in response to a question. Here is a list of the command words used in chemistry and how to respond to them.

Calculate: You need to work out a problem using numbers.

- The problem may be in several stages.
- Look out for the number of marks: this often shows you the number of stages needed in the calculation.
- Always show your working.

Example: Calculate the mass of 11 dm^3 of carbon dioxide. The answer involves: **i.** finding the number of moles of carbon dioxide, using the relationship that 1 mol of a gas occupies 24 dm^3, then **ii.** multiplying moles by the molar mass of carbon dioxide.

Define: You need to be able to write down the main points about a term.

- The only way to do this is by memorising key terms.

Example: Define oxidation in terms of electron transfer. Answer: Oxidation is loss of electrons.

Deduce: You have to work out something from the information given in the question.

Example: Sodium sulfate is Na_2SO_4. Deduce the formula of the sulfate ion. Answer: SO_4^{2-}

Describe: You have to write about a sequence of events, draw a diagram, or state what happens.

- If you are asked to describe observations, remember to state what you see, hear, smell, or feel. You do not need to give the names of the substances.

Example 1: Describe how to obtain sodium chloride crystals from a solution of sodium chloride. Answer: Warm to the crystallisation point and allow to form crystals, then filter off the crystals.

Example 2: Describe your observations when you add acid to an aqueous solution of sodium carbonate. Answer: Bubbles are seen. NOTE: The answer 'carbon dioxide is given off' is not correct.

Example 3: Describe the apparatus used for paper chromatography. When answering, the best way to do this is to include a labelled diagram.

Determine: The answer can't be found directly, but could be obtained from a graph or by calculation.

Example: Determine the value of the gas released after 20 seconds.

Explain: You have to use a particular theory to describe why something happens.

Example: Explain why the volume of a gas increases with increase in temperature. Answer: The particles of gas move faster and get further away from each other (use of kinetic particle theory).

Measure: Take a value directly using suitable measuring instruments.

Example: Measure the temperature of the solution to the nearest °C.

More command words

Predict: You have to make connections between various items of data.

• You often have to extrapolate or interpolate data when answering these questions.

Example: Predict the melting point of potassium (when given the melting points of other Group I elements). The answer involves looking at the melting points of the elements on either side of potassium and choosing a suitable value in between these.

State/Give: Only a short answer is needed.

Example: State/Give the electronic structure of sodium. Answer: 2,8,1.

Suggest: You have to use your general chemical knowledge to write about a situation that is unfamiliar to you. You may need to:

• Think of substances similar to the one that is being asked about.
• Think of general ideas of structure, bonding, electrolysis, redox, rate, or equilibrium to answer the question.

Example: The structure of one type of boron nitride is similar to graphite. Suggest why boron nitride is slippery. (You have to think about the properties of graphite that make it slippery, then repeat these for boron nitride.) Answer: e.g. There are weak forces between the layers, so the layers can slide over each other.

What is meant by: A definition is usually needed.

• The only way to do this is by memorising relevant key terms.

What is meant by the term *isotopes*? Answer: Isotopes are atoms with the same number of protons but different numbers of neutrons.

NOTE: Command words are often combined e.g.

Describe and explain the effect of increasing the temperature on the position of this equilibrium.

1. Underline the command words in each of these questions.
    a. Use the information in the table above to suggest a value for the boiling point of propane.   [2]
    b. Describe how distillation is carried out and give the name of the physical property on which it is based.   [2]
    c. Use this information to deduce the formula of this oxide of tin.   [2]
    d. Describe the process of diffusion and explain this process in terms of the kinetic particle theory.   [2]
    e. Draw a graph of volume of carbon dioxide against time to determine the volume of carbon dioxide formed in the first 30 seconds.   [2]

There is no one correct way of revising. You should find the best way of revising for yourself.

- Don't just read through books or notes and hope that you will remember things.
- Don't leave revision to the last moment. It is best to revise material throughout the year.
- Revision should be active (see unit 22.4).

Here is a check list of things to help you find the best way for you to revise:

- Find the best time of day to revise. Some people revise better in the evening, others in the morning.

  When do I revise best? ....

- Find a time when you will not be disturbed.

  When is this most likely to be? ....

- Find the best conditions needed for you to revise. Some people prefer to revise in absolute silence; others find it useful to have some music in the background.

  Is music in the background really good for you when revising? ....

  ....

- Revise regularly. You may find it useful to revise a topic about a week after you have finished it, to make sure that you have really understood it.

  Do you revise only for exams or tests? ....

- Find the best length of time for each revision session. You may find that several short periods of revision, for example three spells of 20 minutes with breaks in between, are more productive than one longer period of revision.

  What's your best revision span? ....

- Don't imagine that you are revising usefully unless you test yourself from time to time to prove that you are remembering material.

  Do you test yourself? ....

- Make sure that you pay more attention to topics which you find difficult. Don't ignore them!

  What topics in chemistry do you find difficult? ....

  ....

- Do you remember information better in written form or in the form of diagrams?

  ....

Active revision involves you carrying out different sorts of activities and testing yourself to see if your revision has been successful.

- Revising with others, especially with classmates, asking each other questions, is a useful way of helping you remember things.
- Test yourself or get someone else to test you.
- Make a list of key areas that you find difficult and concentrate on these.
- Make up mnemonics like OIL RIG for Oxidation Is Loss of electrons and Reduction Is Gain of electrons.
- Look through the syllabus and text book and list general areas that need revision, e.g. definitions and key terms to learn, e.g. element, isotopes, relative atomic mass. You can test yourself by writing the terms and definitions on a sheet like this:

term	definition
isotopes	Atoms with the same proton number having different numbers of neutrons
element	A substance which contains only one type of atom
empirical formula	A formula which shows the simplest ratio of atoms in a compound

When you think you have learnt the definitions, cover up or fold over the definition side and see if you can remember them.

1. Give three other general topics throughout the syllabus that could be revised in this way.

.... [3]

2. On a separate piece of paper write the two columns as shown below.

properties of acids	result
+ litmus	turns red litmus blue
+ metals	gives salt + hydrogen

Complete the table to give at least four other properties of acids. [4]

3. Make a table similar to the one above to help you revise the properties of hydrocarbons. [12]

Using past papers

- It is useful to look at the wording in past papers to get a feeling for the language used.
- Look not only for the command words but also for the smaller words that instruct you what to do.
- Look at the number of marks given for the question. This often gives you an indication of how many different points you need to include in your answer. For simple questions, however, you may need to write two points to get one mark.
- Underline the key words (see units 21.8 and 22.2).
- Work through calculations and extended exam questions, and use the mark schemes, if available, to see where the marks are awarded.

Mind maps are simplified diagrams which show the main points of a topic in the form of a diagram. They are useful summaries because all the words which get in the way of learning the essential points are removed. They can be made as simple or as complicated as required but it is best to make them simple at the start. Mind maps can also be constructed to show links between different topic areas. An example is shown below.

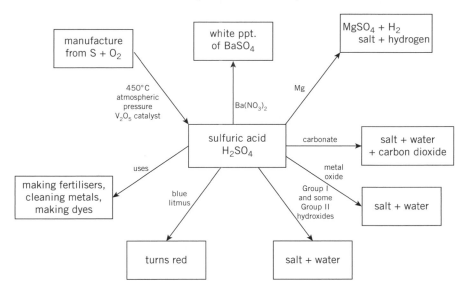

Suggested mind maps which you could make:

electrolysis; acids and bases; methods of purification; hydrocarbons; polymers; properties of the halogens; iron and steel (or properties of metals); equilibrium.

In the space below draw a simple mind map (no more than 7 boxes) for a topic of your choice.

1. Complete the mind map for structure by filling in the gaps **A** to **L**.

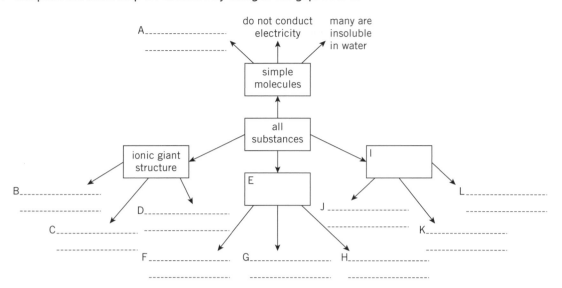

[12]

2. In the space below complete a mind map about rates of reaction.

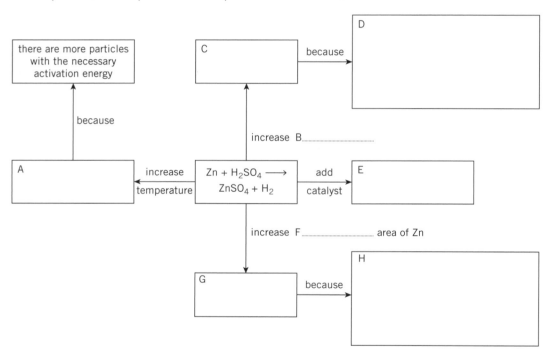

[8]

1. • A small subscript number after an atom or ion refers only to that atom or ion.

   So one formula unit of $CaCl_2$ contains 1 calcium ion and 2 chloride ions and one formula unit of Al_2O_3 contains 2 aluminium ions and 3 oxide ions.

   State how many atoms (or ions) of each type there are in one formula unit of:

   a. Na_2O ....

   b. Mg_3N_2 ....

   c. PCl_3 ....

   d. Al_2O_3 ....

   e. H_2SO_4 .... [5]

2. • When balancing equations, a large number in front of a formula unit multiplies all the way through.
   • The number can refer to atoms (or ions) or moles of atoms (or moles of ions).

   So in $3SO_3$ there are 3×1 sulfur atoms and $3 \times 3 = 9$ oxygen atoms.

   State how many atoms (or ions) of each type there are in the number of formula units shown:

   a. $2H_2S$ ....

   b. $5Mg_3N_2$ ....

   c. $3Al_2S_3$ ....

   d. $2Fe_3O_4$ ....

   e. $3Li_2CO_3$ .... [5]

3. When calculating relative molecular masses (or relative formula masses):
   • Multiply the number of each type of atom by its relative atomic mass and then
   • Add the products together.

   Example: Calculate the relative molecular mass of N_2O_5. (A_r values: N = 14, O = 16)

   $2N = 2 \times 14 = 28$
   $5O = 5 \times 16 = \underline{80}$
   Total $= 108$ Relative molecular mass of $N_2O_5 = 108$

   Calculate the relative molecular mass (or relative formula mass) of these compounds.

   a. Al_2O_3 (A_r values: Al = 27, O = 16)

   .... [1]

   b. Na_2CO_3 (A_r values: C = 12, Na = 23, O = 16)

   .... [1]

   c. $PbSO_4$ (A_r values: Pb = 207, O = 16, S = 32)

   .... [1]

1. • Brackets keep particular groups of atoms together, e.g. (NO_3) for nitrates, (OH) for hydroxides.
   • You must not change the numbers within the brackets.
   • A subscript after a bracket multiplies all through the atoms inside the brackets.

So $Mg(NO_3)_2$ contains 1 Mg ion and 2 NO_3 ions.
In 2NO_3 ions there are 2 N atoms and $2 \times 3 = 6$ O atoms.

And 4$Mg(NO_3)_2$ contain 4 Mg ions, 4×2 N atoms, and 4×6 O atoms.

How many atoms (or ions) of each element are there in one formula unit of:

a. $Sn(SO_4)_2$ ....

b. $(NH_4)_2SO_4$ ....

c. $Ni(ClO_4)_2$ ....

d. $Ba(IO_3)_2$ .... [4]

2. You need to work out the number of atoms correctly in order to calculate the relative formula masses of these compounds.

Calculate the relative formula mass of the compounds in question **1**.

a. $Sn(SO_4)_2$

   Relative formula mass =   [1]

b. $(NH_4)_2SO_4$

   Relative formula mass =   [1]

c. $Ni(ClO_4)_2$

   Relative formula mass =   [1]

d. $Ba(IO_3)_2$

   Relative formula mass =   [1]

3. • Water of crystallisation appears after a dot after the main formula.
   • You add this on separately when calculating formula masses.
   e.g. $CuSO_4.5H_2O$ contains $[(1 \times 64) + (1 \times 32) + (4 \times 16)] + 5 \times (2 + 16) = 250$

Calculate the relative formula mass of $CoCl_2.6H_2O$
(A_r values: Cl = 35.5, Co = 59, H = 1, O = 16)

.... [1]

Learn to rearrange expressions from first principles rather than having to rely on a 'triangle' to help you.

- The idea is that whatever you do to one side of the equation, you do to the other.

  Example: $\text{moles} = \dfrac{\text{mass}}{M_r}$

- To make mass the subject: multiply both sides by M_r (to cancel M_r on the right)

  $\text{moles} \times M_r = \dfrac{\text{mass}}{\cancel{M_r}} \times \cancel{M_r}$ So $\text{moles} \times M_r = \text{mass}$

- To make M_r the subject: divide both sides by *mass* (to cancel mass on the right)

  $\dfrac{\text{moles}}{\text{mass}} = \dfrac{\cancel{\text{mass}}}{M_r \times \cancel{\text{mass}}}$ then turn both sides upside down: $M_r = \dfrac{\text{mass}}{\text{moles}}$

  Now try rearranging these expressions:

1. $\% \text{ yield} = \dfrac{\text{actual yield}}{\text{theoretical yield}} \times 100$

    a. Make actual yield the subject:

        [1]

    b. Make theoretical yield the subject:

        [1]

2. $\text{concentration (in mol/dm}^3) = \dfrac{\text{moles}}{\text{volume (in dm}^3)}.$

    a. Make moles the subject:

        [1]

    b. Make volume (in dm^3) the subject:

        [1]

3. Write the expression for density making mass the subject.

        [1]

1. Sometimes, when you do calculations, the number will come out as, for example, 2.5 E04 on your calculator. The 04 is called the index or power to the base 10. Numbers like this on your calculator are examples of standard form. We write this 2.5×10^4.

   $1 \times 10^1 = 10$

   $1 \times 10^2 = 10 \times 10 = 100$

   $1 \times 10^3 = 10 \times 10 \times 10 = 1000$

   $2.5 \times 10^3 = 2.5 \times 10 \times 10 \times 10 = 2500$

   a. Write 1 000 000 in standard form. .... [1]

   b. Write 7×10^4 in non-standard form. .... [1]

   c. Write 3300 in standard form. .... [1]

   d. Write 3.2×10^3 in non-standard form. .... [1]

2. Very small numbers can also be written in standard form.

   $1 \times 10^{-1} = 0.1$ or 1/10

   $1 \times 10^{-2} = 0.01$ or 1/100

   $1 \times 10^{-3} = 0.001$ or 1/1000

   $2.5 \times 10^{-3} = 2.5 \times 0.001 = 0.0025$

   a. Write 0.00001 in standard form. .... [1]

   b. Write 5×10^{-3} in non-standard form. .... [1]

   c. Write 0.0035 in standard form. .... [1]

   d. Write 2.5×10^{-1} in non-standard form. .... [1]

3. When you multiply numbers in standard form, you simply add the indices (superscripts).

   Example 1: $(2.4 \times 10^3) \times (2 \times 10^2) = 2.4 \times 2 \times 10^{3+2} = 4.8 \times 10^5$

   Example 2: $(5.0 \times 10^{-2}) \times (1.5 \times 10^4) = 5.0 \times 1.5 \times 10^{-2+4} = 7.5 \times 10^2$

   a. What is the product of $(4.0 \times 10^{-3}) \times (3.5 \times 10^2)$? .... [1]

   b. What is the product of $(2.4 \times 10^{-2}) \times (1.5 \times 10^{-3})$? .... [1]

4. When you divide numbers in standard form, you simply subtract the indices.

   Example: $\dfrac{2.4 \times 10^5}{1.2 \times 10^2} = \dfrac{2.4}{1.2} \times 10^{5-2} = 2.0 \times 10^3$

   a. What is the result of $(4.0 \times 10^{-3}) \div (3.5 \times 10^2)$? .... [1]

   b. What is the result of $(2.0 \times 10^2) \div (7.5 \times 10^{-1})$? .... [1]

1. Percentages

In chemistry the result of a smaller number divided by a larger number is multiplied by 100 to get the percentage. You know you've gone wrong in chemical calculations if your percentage yield is greater than 100%!

Example: What is the percentage yield if the actual yield of a product in a reaction is 4.5 g and the theoretical yield is 5.5 g?

$$\% \text{ yield} = \frac{\text{actual yield}}{\text{theoretical yield}} \times 100 = \frac{4.5}{5.5} \times 100 = 82\%$$

Now try these:

a. What is the percentage yield if the actual yield of iron is 20.0 tonnes and the amount expected from the balanced equation is 22.8 tonnes? (Don't forget to multiply by 100!)

% yield =   [1]

b. What is the percentage purity of compound **A** if the mass of impure substance in 3.45 kg of **A** is 0.12 kg? (Note that it's the percentage purity not the percentage impurity that is needed!)

% purity =   [2]

Volumes and areas

2. The diagram shows how to calculate the area of one side of a cube and the volume of a cube.

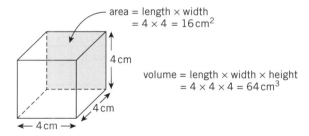

area = length × width
= 4 × 4 = 16 cm²

volume = length × width × height
= 4 × 4 × 4 = 64 cm³

Look at the diagram of the cube above.

a. i. How many sides does the cube have? .... [1]

ii. What is the total surface area of the cube? .... [1]

b. Calculate the volume of a cube that has a side of 5 cm³.

   .... [1]

1. When doing chemical calculations, it is important that we give the answer to the correct number of significant figures and round figures correctly.

Significant figures

236.38 has 5 significant figures

32.4 has 3 significant figures

0.0067 has two significant figures (zeros before a number **are not** significant figures)

0.0300 has 3 significant figures (zeros after a number after a decimal point **are** significant figures)

Rounding up

2.3<u>66</u> rounded to two significant figures is 2.4

2.5<u>57</u> rounded to two significant figures is 2.6

2.3<u>46</u> rounded to two significant figures is 2.3

You can see that when rounding if the next figure along is 5 or above, then the figure to be rounded goes up by 1.

a. Round these values to 3 significant figures:

i. 4.357 ....

ii. 0.08732 ....

iii. 137.2 ....

iv. 0.005498 .... [4]

b. Round these values to 2 significant figures:

i. 436 ....

ii. 3.447 ....

iii. 56.79 ....

iv. 0.00545 .... [4]

2. When performing a calculation in several stages **do not round between the steps**. You should only round at the end. You should round to the same number of significant figures as the data in the question.
   To see the effect of rounding in the middle of a calculation, work through this example.

   When 1 mole of pentane is burned in excess air, 5 moles of carbon dioxide are formed. Calculate the volume of carbon dioxide when 6.67 g of pentane burn.

calculation	answer when keeping the figures in your calculator	value when rounding
moles pentane $= \dfrac{6.67}{72.0}$	....	round to 1 significant figure ....
multiply by 5 (because 5 moles of carbon dioxide are formed from 1 mole of pentane)	....	round to 1 significant figure ....
multiply by 24 to get dm³ of carbon dioxide	answer to 3 significant figures	answer to 3 significant figures ....

[6]

- We draw bar charts in chemistry to compare values of different physical properties such as melting points, temperatures changes, or strength. We can also use them to compare other quantities or numbers.
- The bars are shown as rectangles of the same width, which can be vertical or horizontal.
- The bars can be separated from each other or join each other.

The diagram shows two bar charts representing the densities of different elements.

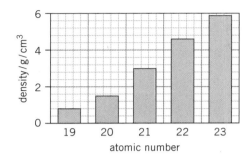
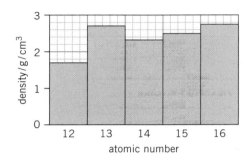

1. The table shows the temperature change when equal amounts of different chlorides **A** to **E** dissolve in the same volume of water.

chloride	A	B	C	D	E
temperature change / °C	+2.5	+4.6	+3.2	−1.5	−0.4

Draw a bar chart to show these results.
- Make sure that you use the full range of the graph paper and label the axes.
- Make sure that you take the negative values into account: the 0 on the *y*-axis needs to be a little way up the graph paper.

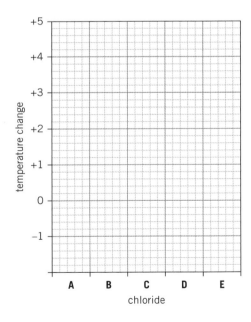

[4]

1. When drawing graphs remember:
   - The axes should be fully labelled and include units.
   - Use as much of the graph paper as possible.
   - Use an × to plot the points rather than + or •, which cannot be so easily seen.
   - If it looks as though the line will form a curve, do not use a ruler to join the points to each other.
   - Draw the line of best fit (with equal numbers of points each side of the line if necessary).
   - Ignore any points which do not fit in with the general trend of the line (anomalous points).

   What is wrong with each of these graphs?

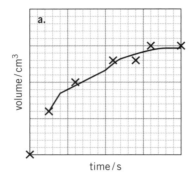
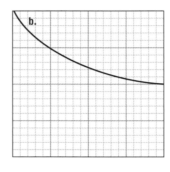
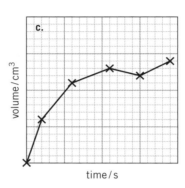

   a. ....

   ....

   .... [3]

   b. ....

   ....

   .... [3]

   c. ....

   .... [2]

2. Extrapolation and interpolation.

   The graph on the right shows how to extrapolate and interpolate values. Always make sure that you draw lines as shown to the values that are asked for.

   Determine the volume of gas released in the first:

   a. 2.4 minutes. .... [1]

   b. 5 minutes .... [1]

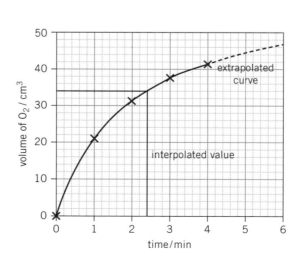

1. The table shows how the electrical conductivity of a solution changes as an acid is added to an alkali.

volume of acid / cm³	0	2	3	4	5	6	7	8
conductivity / ohms⁻¹m⁻¹	2.8	1.8	1.3	0.8	0.7	1.0	1.2	1.5

   a. On the grid below plot the points using the data in the table. [3]

   b. Draw two straight lines to connect these data points that intersect. Label this intersection point P. [2]

   c. Determine the volume of acid added at point P. .... [1]

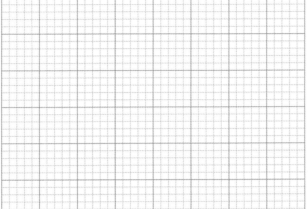

2. The table shows how the volume of gas changes when magnesium reacts with hydrochloric acid. Plot a graph of these results on the grid below. [4]

time / s	0	10	20	30	40	50	60	70
volume of gas / cm³	0	24	39	47	52	54	55	56

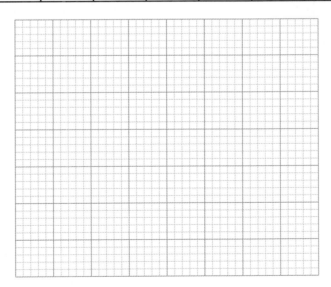

1. Look at the graph below.

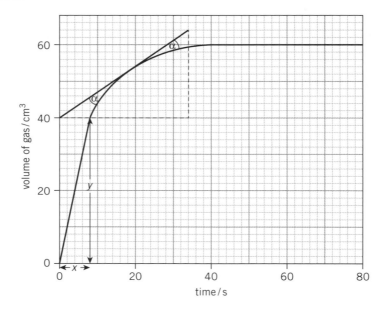

- We can find the initial rate of reaction from a graph by taking the rise, *y*, over the run, *x*, of the gradient (slope). This is 40 cm³ ÷ 8 s. So the rate is 5 cm³/s.

- We can find the rate at any point on a curve by drawing a tangent to the curve (see the graph above). Note that the angles α should be equal.

In this case the gradient is $\dfrac{64 - 40}{34 - 0} = 0.71$ cm³/s

a. The table shows how the mass of a product in a reaction increases with time.

time / s	0	20	40	60	80	100	120	140
mass / g	0	0.08	0.16	0.24	0.30	0.34	0.37	0.40

Plot a graph using these results.

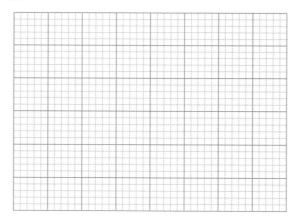

[4]

b. Calculate the initial rate of reaction over the first 40 seconds.

.... [2]

c. Why would the rate calculated over the first 100 seconds be an average rate?

.... [1]

1. Name the most important thing that you wear in the laboratory.

.... [1]

2. Hydrogen sulfide is a poisonous gas that can be prepared by warming dilute sulfuric acid with iron(II) sulfide. Hydrogen sulfide is heavier than air.

   What precautions would you take when carrying out this experiment in the laboratory?

.... [2]

3. What is the meaning of each of these hazard symbols?

A ....   B ....   C ....

D ....   E ....   [5]

4. Describe the hazards associated with each of these chemicals.

   a. Ethanol .... [1]

   b. Concentrated potassium hydroxide .... [1]

   c. Chlorine .... [1]

5. The apparatus below was used to prepare hydrogen chloride gas. Identify and explain three errors.

....

....

....

....

.... [6]

1. State five things you need to include when you plan an experiment.

    ....

    ....

    ....

    ....

    .... [5]

    When planning an experiment, you have to think about what things you can change (the variables) and what things you can measure.

    *   The variable that you choose the values for is the *independent variable*, e.g. 10 s, 20 s, 30 s, and so on.
    *   The variable that you measure at each of these values is the *dependent variable*, e.g. 15 cm³ gas at 10 s, 28 cm³ gas at 20 s, and so on.
    *   All the other values that have to be kept constant to make the experiment a fair test are the *control variables*.

    Identify the three types of variable in the experiments in questions 2 and 3.

2. Investigation into the effect of temperature on the rate of reaction of hydrochloric acid with calcium carbonate by measuring the volume of carbon dioxide released after 10 s.

    a. Independent variable: .... [1]

    b. Dependent variable: .... [1]

    c. Control variables: ....

    .... [2]

3. The energy released on burning different fuels was compared by measuring the temperature rise of the water in the copper can when 1 g of fuel was burnt.

    a. Independent variable: .... [1]

    b. Dependent variable: .... [1]

    c. Control variables: ....

    .... [2]

1. Accurate results are close to the true value. State two things that will help you get accurate results.

    ....

    .... [2]

2. A student added aqueous 2.0 mol/dm³ hydrochloric acid from a burette to 20 cm³ of a solution of sodium hydroxide in a flask. After each addition of acid, the temperature of the mixture in the flask was measured. Complete the table below by taking the thermometer and burette readings.

burette diagram	total volume of acid added / cm³	thermometer diagram	maximum temperature / °C
0 / 1		25 / 20	
4 / 5		25 / 20	
8 / 9		30 / 25	
12 / 13		30 / 25	

[4]

3. What volumes, in cm³, are shown on the gas syringes **S** and **T**?

    S 10 20 30 40 —part of plunger   **T** 10 20 30 40 —part of plunger

    .... cm³   .... cm³   [2]

1. A flask containing 0.25 mol/dm³ hydrochloric acid was placed on a digital balance and excess calcium carbonate was added. The balance was immediately set to zero and the readings of decrease in mass were taken every 10 seconds. The total decrease in mass at each 10-second interval is given below.

   start 0.0 g, 0.50 g, 0.89 g, 1.24 g, 1.44 g, 1.56 g, 1.68 g, 1.79 g, 1.82 g, 1.82 g

   The results from a repeat experiment are:

   start 0.0 g, 0.72 g, 1.04 g, 1.24 g, 1.44 g, 1.52 g, 1.62 g, 1.74 g, 1.78 g, 1.78 g

   Draw a suitable table to display these results and obtain an average value of the change in mass.

   [4]

2. The apparatus shown below can be used to show how reaction rate changes with temperature for the reaction between sodium thiosulfate and hydrochloric acid.

   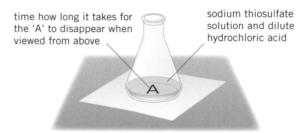

   time how long it takes for the 'A' to disappear when viewed from above

   sodium thiosulfate solution and dilute hydrochloric acid

   As the reaction proceeds, a precipitate of sulfur is made, which gradually makes the letter 'A' disappear.

   The rate is proportional to $\dfrac{1}{\text{time taken for the letter 'A' to disappear}}$.

   Draw the headings of a suitable table to enable you to record the results of the experiment at different temperatures.

   [2]

- Numerical data with dependent and independent variables that are both numbers are best displayed on a line graph.
- The independent variable should be the *x*-axis (horizontal axis) of the graph.
- The dependent variable should be the *y*-axis (vertical axis) of the graph.

1. Identify the independent variable and the dependent variable in these two cases.

    a. The electrical conductivity of a solution is measured at 2-minute intervals.

    Independent variable ....

    Dependent variable .... [2]

    b. The rate of reaction of hydrochloric acid with sodium thiosulfate is studied using four different concentrations of acid. At each concentration, the time taken for a precipitate to make a letter M 'disappear' is recorded.

    Independent variable ....

    Dependent variable .... [2]

2. The rate of hydrolysis of compound **A** can be deduced from measurements of electrical conductivity of the reaction mixture. The results are shown in the table.

initial concentration of **A** / mol/dm³	relative electrical conductivity					
	0 min	1 min	2 min	3 min	3.5 min	4 min
1.4	0	2.5	5.0	7.5	9.2	10.0
2.2	0	3.8	7.5	11.2	13.0	15.0
3.2	0	6.0	12.0	16.0	21.1	24.0
4.0	0	7.5	15.0	22.5	26.0	30.0

    a. On the grid below plot a graph of these results.

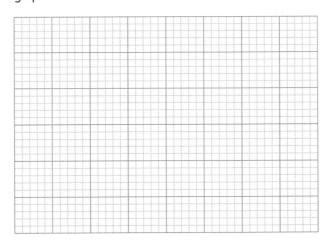

[5]

    b. Which point is anomalous? Explain why.

    ....

    .... [2]

When you have recorded the data, you need to display it in the best way.

- First order your data in a table (see unit 24.4).
- Then decide the best way of displaying it, using graphs or charts.
- Numerical data with dependent and independent variables that are both numbers are best displayed on a line graph (see unit 24.5).
- Data in which one of the variables is not a number are best displayed as a bar chart.
- Sometimes conclusions have to be reached from the data in the table.

1. Hydrogen peroxide, H_2O_2, reacts with acidified iodide ions, I^-. The table shows the relative rate of reaction using different concentrations of hydrogen peroxide and iodide ions (experiments **A** to **E**).

	concentration of H_2O_2 / mol/dm³	concentration of I^- / mol/dm³	relative rate of reaction
A	0.01	0.01	1.75
B	0.02	0.01	3.50
C	0.02	0.02	7.00
D	0.02	0.03	10.50
E	0.03	0.01	5.25

a. i. Which three experiments would you compare if you wanted to find how the concentration of hydrogen peroxide affects the rate of reaction?

.... [1]

ii. How does the concentration of hydrogen peroxide affect the rate of the reaction?

.... [2]

b. i. Which three experiments would you compare if you wanted to find how the concentration of iodide ions affects the rate of reaction?

.... [1]

ii. How does the concentration of iodide ions affect the rate of the reaction?

.... [2]

c. Why is the experiment using three different concentrations of hydrogen peroxide and three different concentrations of iodide ions the minimum amount of data that is needed to draw the correct conclusions about the effect of the concentrations on rate of reaction?

....

.... [2]

2. Look at the graph you plotted in unit 24.5. How could you use this graph to show that the rate is directly proportional to the initial concentration of **A**?

....

....

.... [3]

Conclusions

- Make sure that you describe the patterns in the data accurately, stating the direction of **change** in both the dependent and independent variable, e.g. the rate increases as the concentration increases. (The statement 'Rate increases with concentration' is less accurate because we do not know if concentration is increasing or decreasing.)
- Describe graphs accurately. The line is only proportional if it is a straight upward-sloping line going through the 0-0 point.
- If there is a change in the gradient of a line or curve, make a comment about this, e.g. it is a downward-sloping curve whose gradient decreases with time.

1. Describe the patterns shown by graphs **A**, **B**, and **C** as accurately as possible.

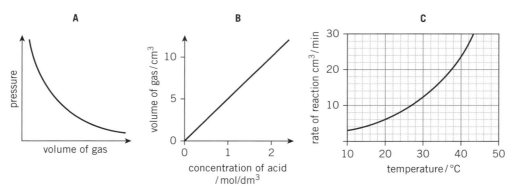

a. Graph **A** ....

.... [2]

b. Graph **B** ....

.... [2]

c. Graph **C** ....

.... [2]

Evaluation

When evaluating an experiment we need to think about:

- The accuracy of the measuring apparatus used. The accuracy of the whole experiment depends on the accuracy of the least accurate measuring instrument.
- The number of times the experiment is repeated. If similar results are obtained when the experiment is repeated several times, we can be sure that the results are consistent.
- How easy it is to control the variables. If it is not easy to control a variable, it is not a fair test. For example, if you try to keep a fixed temperature in a beaker of water using a Bunsen burner, you are unlikely to succeed. So use a fixed-temperature water bath instead.

Improvement

We can improve our results by:

- Using more accurate measuring apparatus, e.g. a thermometer with 0.1 °C scale divisions instead of 1 °C scale divisions.
- Repeating the results as many times as possible in the time allowed.
- Making sure that the variables are carefully controlled.

Many salts increase in solubility as the temperature increases. Plan an experiment to see if this is true using the salt potassium chloride as an example. This exercise guides you through the stages in this experiment.

1. Planning:

   a. What apparatus do you need?

   ....

   ....

   .... [5]

   b. What are the independent and dependent variables?

   ....

   .... [2]

   c. What will you keep constant?

   .... [2]

2. Carrying out:
   Describe how you will carry out the experiment. Give possible volumes and masses of the substances used.

   ....

   ....

   ....

   ....

   .... [5]

3. Evaluation:

   a. Explain why it might be difficult to control the temperature.

   ....

   .... [2]

   b. Suggest improvements that you could make to the experiment.

   ....

   .... [2]

   c. Which measurement is most likely to have the greatest percentage error?
      Explain your answer.

   ....

   .... [2]

138

Hydrogen peroxide at a concentration of 2 mol/dm³ can be decomposed by small portions of metal oxides such as manganese(IV) oxide, copper(I) oxide, and lead(IV) oxide.

$$2H_2O_2(aq) \rightarrow 2H_2O(l) + O_2(g)$$

Plan an experiment to compare the ability of each of these compounds to decompose hydrogen peroxide.

1. Planning:

    a. What apparatus do you need?

        ....

        ....

        .... [5]

    b. What are the independent and dependent variables?

        ....

        .... [2]

    c. What will you keep constant?

        ....

        .... [3]

2. Carrying out:
    Describe how you will carry out the experiment. Give possible volumes and masses of the substances used.

    ....

    ....

    ....

    ....

    .... [5]

3. Evaluation:

    a. The reaction can be quite fast. Explain any problems associated with this.

        ....

        .... [2]

    b. Suggest improvements that you could make to the experiment.

        ....

        ....

        .... [2]

139

Before the test

- Make sure you know how to use apparatus such as a burette, volumetric pipette, and dropping pipette.
- Make sure that you can take readings correctly from a burette, measuring cylinder, and mass balance.
- Make sure that you have plenty of practice at plotting graphs and drawing bar charts.
- Study past test papers carefully to see the types of questions that are asked.

Test questions are of two main types:

- Questions involving a titration, measuring heat changes, following a reaction to deduce rates, electrolysis, or determining solubility.
- Questions on qualitative analysis (tests for ions and gases). It helps if you know these thoroughly even if you have access to a data sheet. After all, you have to know them for the theory papers!

What you may be asked to do:

- Read apparatus scales with precision.
- Process data by constructing tables and graphs.
- Interpolate and extrapolate data (especially graphical data).
- Do simple calculations with the aid of a calculator.
- Draw conclusions from the data.
- Evaluate experiments and data.
- Identify sources of error and suggest improvements to an experiment.

During the test

- Read the questions carefully to make sure you know exactly what to do.
- Keep calm and divide your time appropriately between the (two) experiments.
- If you make a mistake in your answer, delete the incorrect answer with a single line and rewrite the answer so that the marker is aware of where it is.
- If you make a practical mistake start the exercise again if you have enough material.
- Record the temperature to nearest °C.
- Use the amount of materials given in the question. For qualitative analysis questions use the minimum amount of material for the experiment. About 1 cm depth of solution is usually enough.
- Don't throw any substance away (you may need it if you make a mistake or need it for an additional test).
- Label your test tubes. This is especially important when you need a solution for an additional test or there are several different colourless solutions.
- Don't use the same dropping pipette for different solutions. If necessary, wash the pipette out with distilled water and make sure that it is fairly dry before using it for a different solution.
- When adding a solution from a dropping pipette to another solution in a test tube, add the solution slowly so that you can observe any gradual changes.

Writing your observations

- Make sure that you write your observations as you go along, not after doing several operations.
- Make sure that you include any colour changes.
- Make sure that the states are included, e.g. precipitate formed / bubbles of gas.

Introduction

This could be done at home or in the school laboratory.

- Hard water does not lather well with soap.
- Soft water lathers well with soap.
- Permanent hardness in water cannot be removed by boiling.
- Temporary hardness in water can be removed by boiling.

Purpose of the experiments

To find the volume of soap solution (or washing-up liquid) needed to form a permanent lather with different samples of water.

Sources of water

- Distilled water.
- Temporary hard water (bubble carbon dioxide through limewater until the white precipitate has disappeared).
- Permanent hard water (add some hydrated calcium sulfate to distilled water then filter).
- Natural sources of water, e.g. tap water, rainwater, seawater.

Carrying out the experiments

1. There are two ways in which the experiment can be carried out.
   a. Add the water sample to a flask and then see how many drops of soap solution (added from a burette or pipette) are needed to form a lather on shaking that does not disappear after leaving for a minute.
   b. Adding a certain number of drops of soap solution to a tube of water, shaking, and measuring the height of the lather formed.
2. Make a list of all the equipment that you need including safety equipment/clothing.
3. What do you need to vary and what do you need to keep constant?

Analysing the results

- Draw a table of results for the number of drops (or height of lather) with different types of water including boiled hard water (temporary and permanent).
- Repeat your experiments to get consistent results.
- Suggest how you could improve your experiments.

Conclusions

1. Which samples of water are hard and which are soft?
2. Classify tap water, rainwater, seawater, and other sources of water you have analysed as hard or soft.
3. Which samples contain temporary hardness and which contain permanent hardness?

Introduction

This could be done at home or in the school laboratory.

- The labels on packets of food usually state the amount of energy they contain in kilojoules or kilocalories. Make a list of these values for the foods you choose.
- When dry foods are burnt, they release energy. The reaction is exothermic.
- The carbohydrates, fats, and proteins in the food burn to form carbon dioxide and water.

Purpose of the experiments

To compare the energy released by different foods.

Sources of foods

- The foods should be dry. You can dry wet foods in an oven but don't let them char (go black).
- Make a list of the energy values for the foods you choose by looking on the sides of the packets.
- Crisps, bread, nuts, rice are good sources.
- Dried meats, beans, and cheese could also be used.
- It is possible to burn cooking oils if you use a cotton or string wick.

Carrying out the experiments

1. There are two ways in which the experiment can be carried out.
   a. Burning different foods of known mass on the end of a large needle. The burning foods heat a known volume of water in a test tube or beaker.
   b. Burning the food on a tin lid beneath a beaker or tin of water. This is more useful for fats and oils.
2. Make a list of all the equipment that you need including safety equipment / clothing.
3. What do you need to vary and what do you need to keep constant?
4. You could also investigate the relationship between the mass of a particular food burnt and the temperature rise.

Analysing the results

- Draw a table of results for the temperature rise on burning a known amount of food material using a fixed volume of water.
- Repeat your experiments to get consistent results.
- Suggest how you could improve your experiments.

Conclusions

1. Calculate the energy released in kJ per gram of food by using the relationship:
   energy released (Joules) = mass of water (g) × 4.18 × temperature rise (°C)
2. Which foods released the most energy per gram?
3. Compare the energy values you obtained with the energy values on the labels of the foods you used. Were they in the same order of energy as the results of your experiments? If not, suggest why not.
4. Suggest reasons why your experiment may not be a fair test.

Introduction

This is best done in the laboratory.

- Many compounds are much more soluble in water at higher temperatures than lower temperatures. Others do not show much difference in solubility as the temperature increases.

Purpose of the experiments

To find how the solubility of different compounds changes with temperature.

Suggested compounds to use

Potassium nitrate, sodium nitrate, potassium chloride, and sodium chloride.

Carrying out the experiments

1. a. Heat some water (4 or 5 cm³) with the solute until the solute dissolves.

    b. The solution is then cooled using the apparatus shown below until crystallisation occurs. The temperature of crystallisation is recorded.

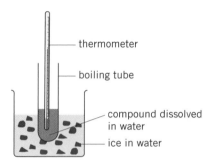

thermometer

boiling tube

compound dissolved in water

ice in water

    c. Then add more water (not more than 2 cm³) to the solution in the boiling tube and heat to dissolve. The temperature when crystals appear is recorded.

    d. Repeat step **c.** several times.

2. Make a list of all the equipment that you need including safety equipment/ clothing.

3. What do you need to vary and what do you need to keep constant?

Analysing the results

- For each salt draw a table of results for the temperature at which crystallisation occurs.
- Repeat your experiments to get consistent results.
- Suggest how you could improve your experiments.

Conclusions

1. Which compounds show the greatest difference in crystallisation temperature?

2. Compare your results with tables showing the solubility of each of these compounds at different temperatures.

3. Suggest reasons why your experiment may not be a fair test.

Introduction

This should be done in the laboratory.

- When carbonates react with acids carbon dioxide is released.
- The mass of the reaction mixture decreases as the reaction proceeds.
- The diagram below shows some of the apparatus that can be used to follow the rate of this reaction.

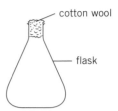

cotton wool

flask

Purpose of the experiments

To find the percentage of carbon dioxide and hence the percentage by mass of carbon in different carbonates.

Suggested carbonates to use

Sodium carbonate, sodium hydrogencarbonate, calcium carbonate, copper(II) carbonate, barium carbonate.

Carrying out the experiments

1. The hydrochloric acid needs to be in excess. Why?
2. Deduce the amounts of hydrochloric acid and calcium carbonate you need.
3. Make a list of all the equipment that you need including safety equipment / clothing.
4. What do you need to vary and what do you need to keep constant?

Analysing the results

- Draw up a table of results for the mass or volume of carbon dioxide given off for each carbonate.
- Repeat your experiments to get consistent results.
- Suggest how you could improve your experiments.
- Calculate the mass of carbon released in each experiment.
- Calculate the percentage by mass of the carbon in the carbonate.

Conclusions

1. Put the carbonates in order of their percentage composition carbon by mass.
2. The CO_3^{2-} ion is common to all the carbonates. So why are the percentage compositions different?
3. Suggest reasons why your experiment may not be a fair test.

1. Phosphorus is an element in Group V of the Periodic Table.

    a. Deduce the electron arrangement of an atom of phosphorus.

    .... [1]

    b. An isotope of phosphorus has 15 protons and 31 nucleons.

    Deduce the number of neutrons in this isotope of phosphorus.

    .... [1]

    c. Phosphorus has a simple molecular structure.

    Describe two physical properties of phosphorus.

    ....

    .... [2]

    d. Phosphorus burns in excess oxygen to form an oxide with the formula P_2O_5.

    Write a balanced equation for this reaction.

    .... [2]

    e. P_2O_5 reacts with sodium hydroxide to form sodium phosphate, Na_3PO_4.

    Deduce the formula of the phosphate ion.

    .... [1]

    f. Phosphate ions are present in many fertilisers.

    Name another anion that is present in most fertilisers.

    .... [1]

    g. Explain why farmers spread fertilisers on the soil where crop plants are grown.

    ....

    .... [2]

    h. Draw the electronic structure of phosphine, PH_3. Show only the outer shell electrons.

    [2]

    Total = 12

2. The structure of allyl alcohol is shown below.

$$CH_2=CH-CH_2-OH$$

a. What feature of allyl alcohol shows that it is an unsaturated compound?

.... [1]

b. Describe a test for an unsaturated compound.

Test ....

Result .... [2]

c. Allyl alcohol can be reduced by hydrogen in a similar way to ethene.

   i. Explain the term *reduction* in terms of electron transfer.

   .... [1]

   ii. State the conditions needed for this reduction.

   ....

   .... [3]

   iii. Give the full structural formula of the compound formed by this reduction. Show all atoms and all bonds.

   [2]

d. Compounds with structures similar to allyl alcohol are found in green onion leaves.

   i. Suggest how you could make a solution of the green pigments from onion leaves.

   ....

   .... [2]

   ii. Several green pigments are present in onion leaves. State the name of the method you would use to separate these pigments from each other.

   .... [1]

Total = 12

3. The structure of caesium chloride is shown below.

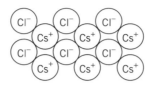

a. Deduce the simplest formula for caesium chloride.

.... [1]

b. Explain in terms of structure and bonding why caesium chloride has a high melting point.

....

.... [2]

c. Explain why aqueous caesium chloride conducts electricity.

.... [1]

d. Molten caesium chloride is electrolysed using graphite electrodes.

   i.   Give two reasons why graphite electrodes are used.

   ....

   .... [2]

   ii. Write ionic half-equations (ion electron equations) for the reactions at:

   the anode ....

   the cathode .... [3]

e. Caesium chloride is formed when caesium burns in chlorine.

$$2Cs \ + \ Cl_2 \ \rightarrow \ 2CsCl$$

When 5.32 g of caesium are burnt in excess chlorine, 6.4 g of caesium chloride are formed. Calculate the percentage yield of caesium chloride.

[3]

Total = 12

4. A student investigated the reaction at r.t.p. between 0.05 g magnesium ribbon and excess hydrochloric acid of concentration 2.0 mol/dm³.

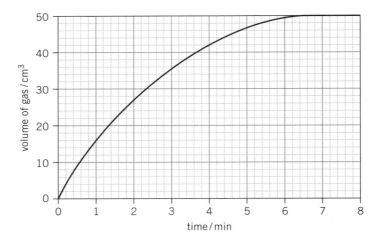

a. At what time was the reaction just complete?

.... [1]

b. i. Deduce the volume of hydrogen released during the first minute of the reaction.

.... [1]

ii. Deduce the average rate of reaction during the first two minutes.

.... [1]

c. The experiment was repeated at r.t.p. using hydrochloric acid of concentration 2.5 mol/dm³. On the grid above draw a line to show how the volume of hydrogen released changes with time. [2]

d. Explain, using the collision theory, why changing the concentration of acid affects the rate of reaction.

....

....

.... [2]

e. The experiment was repeated using 2 mol/dm³ hydrochloric acid and 0.05 g magnesium powder. Would the reaction be faster or slower? Explain your answer.

.... [2]

Total = 9

5. When 1 mole of calcium carbonate is heated, 1 mole of calcium oxide and 1 mole of carbon dioxide are formed.

   a. What type of reaction is this? Put a ring around **two** of the words below.

       addition   catalysed   decomposition   endothermic

           exothermic   oxidation   reduction   [2]

   b. Describe a test for carbon dioxide.

     Test ....

     Result .... [2]

   c. The table shows the mass of carbon dioxide formed when calcium carbonate is heated for 5 minutes at different temperatures. The same mass of calcium carbonate was used in each experiment.

temperature / °C	500	700	800	900	950	1000
mass / g	0.0	0.4	1.8	3.5	3.7	3.8

     i. On the grid below draw a graph of these results.

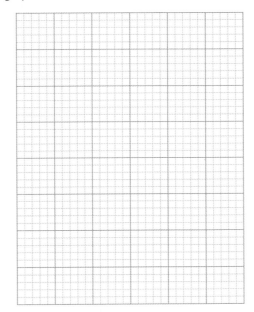

                                                                 [3]

     ii. Use your graph to help you calculate the volume of carbon dioxide formed when calcium carbonate is heated for 5 minutes at 850 °C.

                                                                      [3]

                                                                    Total = 10

6. The diagram shows the preparation of ammonia by heating ammonium sulfate with concentrated sodium hydroxide.

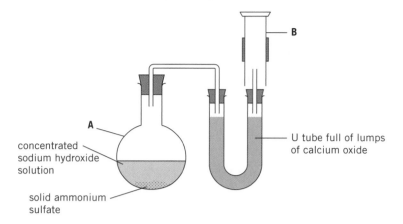

concentrated
sodium hydroxide
solution

A

B

U tube full of lumps
of calcium oxide

solid ammonium
sulfate

a. i. On the diagram above, show where heat is applied. [1]

ii. State the names of the pieces of apparatus labelled **A** and **B**.

A ....

B .... [2]

iii. What is the purpose of the calcium oxide?

.... [1]

iv. Explain how you can show when **B** is full of ammonia.

....

.... [2]

b. Complete the equation for the reaction.

$(NH_4)_2SO_4$ + ... $\rightarrow$ .... + ... + H_2O [2]

c. Hydrazine, H_2N-NH_2, like ammonia, contains hydrogen and nitrogen.

Draw the electronic structure of a molecule of hydrazine. Show only the electrons in the outer shells.

[2]

Total = 10

7. The structure of lactic acid is shown below.

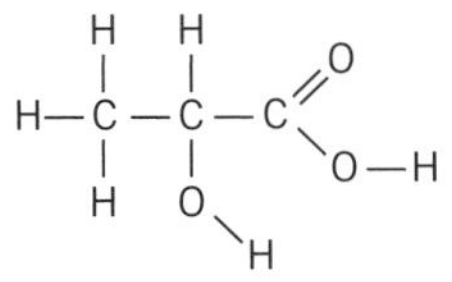

a. On the structure above put a ring around the alcohol functional group. [1]

b. Lactic acid can be made by fermenting the sugar lactose.

   i. State the three different types of atom present in sugars.

   .... [2]

   ii. Give the name of another compound that can be made by fermentation.

   .... [1]

c. Calcium carbonate neutralises lactic acid. Complete the word equation for this reaction.

   lactic acid   +   calcium carbonate   →   calcium lactate   + ...   + ...   [2]

d. Calcium lactate is insoluble in water. Suggest how you could separate calcium lactate from a mixture of calcium lactate and aqueous salts.

   .... [1]

e. The simplified structure of the polymer of lactic acid is shown below.

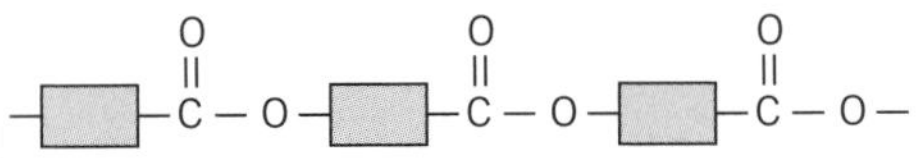

   i. State the name of the linkage group.

   .... [1]

   ii. Explain why this is not an example of addition polymerisation.

   ....
   .... [2]

f. Lactic acid is oxidised to ethanoic acid by acidified potassium manganate(VII). What colour change would you observe when excess lactic acid is added to acidified potassium manganate(VII)?

   .... to .... [2]

   Total = 12

8. The table shows some physical properties of four noble gases.

gas	melting point / °C	boiling point / °C	density at r.t.p. / g/dm³	atomic radius / nm
helium	−272	−269	0.18	0.050
neon	−248	−246	0.90	0.065
argon	−189	−186	1.78	
krypton	−157	−152	3.74	0.110

a. i. The density of air is 1.20 g/dm³. Which of these gases could be used to fill a toy balloon to float in air?

.... [1]

ii. Deduce the atomic radius of argon.

.... [1]

iii. What is the state of krypton at −118 °C? Explain your answer.

.... [2]

iv. Describe the trend in boiling point down the Group.

.... [1]

b. Xenon tetrafluoride, XeF_4, reacts with potassium iodide.

$$XeF_4 \;+\; 4KI \;\rightarrow\; Xe \;+\; 2I_2 \;+\; 4KF$$

i. Potassium salts are colourless. What is the final colour of the reaction mixture?

.... [1]

ii. Explain why potassium iodide is acting as a reducing agent in this reaction.

....

.... [2]

iii. Calculate the maximum volume of xenon formed when 8.28 g of xenon tetrafluoride reacts with excess potassium iodide.

[3]

Total = 11

9. Nitrogen dioxide, NO_2, is a brown gas that pollutes the atmosphere.

   a. i. Give one source of nitrogen dioxide in the atmosphere.

      .... [1]

      ii. Describe one effect of nitrogen dioxide on the environment.

      .... [1]

      iii. Nitrogen dioxide is a gas at r.t.p. Describe the proximity (closeness) and motion of the particles in nitrogen dioxide at r.t.p.

      ....

      .... [2]

   b. The colourless gas dinitrogen tetroxide, N_2O_4, forms an equilibrium mixture with nitrogen dioxide.

      $$N_2O_4(g) \rightleftharpoons 2NO_2(g)$$

      i. Describe and explain what you would observe when the pressure on this equilibrium mixture is increased.

      ....

      ....

      .... [3]

      ii. Calculate the relative molecular mass of:

      nitrogen dioxide ....

      dinitrogen tetroxide .... [2]

      iii. At 55 °C the average relative molecular mass of the equilibrium mixture is 61.0 but at 140 °C, the average relative molecular mass is 46.0.

      Explain how this shows that the reaction is endothermic.

      ....

      ....

      .... [2]

   c. At temperatures above 150 °C nitrogen dioxide decomposes to nitrogen(II) oxide and oxygen. Write a symbol equation for this reaction.

      .... [2]

      Total = 13

10. 25 cm^3 of aqueous potassium hydroxide was placed in a flask. A few drops of an acid–base indicator were then added. The solution was neutralised by 12.5 cm^3 of 0.2 mol/dm^3 sulfuric acid added from a burette.

$$2KOH(aq) \quad + \quad H_2SO_4(aq) \quad \longrightarrow \quad K_2SO_4(aq) \quad + \quad 2H_2O(l)$$

a. Suggest a suitable indicator that could be used in this reaction.

.... [1]

b. Give the name of the salt formed in this reaction.

.... [1]

c. Calculate:

   i. The number of moles of sulfuric acid added from the burette.

[1]

   ii. The number of moles of potassium hydroxide in the flask.

.... [1]

   iii. The concentration of potassium hydroxide in the flask in mol/dm^3.

[1]

d. Write the simplest ionic equation for this reaction.

.... [1]

e. Sulfuric acid catalyses the reaction between ethanol and butanoic acid.

Draw the full structural formula for the ester formed in this reaction, showing all atoms and all bonds.

[2]

Total = 8

Glossary

Acid: A proton donor.

Acid–base indicator: A coloured compound or mixture of coloured compounds that changes colour over a specific pH.

Acidic oxide: An oxide that react with alkalis to form a salt and water.

Acid rain: Rain that has a pH below about pH 5 due to the reaction of rainwater with acidic gases.

Activation energy: The minimum amount of energy particles must have to react when they collide.

Addition polymerisation: Polymerisation of monomers containing a C=C double bond to form a polymer when no other compound is formed.

Addition reaction: A reaction in which a single product is formed from two or more reactant molecules and no other product is made.

Alcohols: Organic compounds with branched or unbranched chains containing the −OH functional group.

Alkali: A base that is soluble in water.

Alkanes: Saturated hydrocarbons with the general formula C_nH_{2n+2}.

Alkenes: Hydrocarbons containing at least one C=C double bond.

Alkyl group: A group formed by the removal of a hydrogen atom from an alkane, e.g. CH_3-, C_2H_5-.

Alloy: A mixture, within a metallic lattice, of two or more metals or a mixture of one or more metals with a non-metal.

Amphoteric oxide: An oxide that reacts with both acids and alkalis.

Anions: Negative ions.

Anode: The positive electrode.

Atom: The smallest particle that cannot be broken down by chemical means.

Atomic number: The number of protons in the nucleus of an atom.

Base: A proton acceptor.

Basic oxide: An oxide that reacts with acids to form a salt and water.

Brownian motion: The random bombardment of molecules on small suspended particles leading to a random irregular motion of the suspended particles.

Carboxylic acids: A homologous series of organic compounds with the −COOH group.

Catalyst: A substance that speeds up a chemical reaction but is unchanged at the end of the reaction.

Catalytic converter: Part added to vehicle to reduce the emissions of carbon monoxide and nitrogen oxides from exhausts of petrol engines.

Cathode: The negative electrode.

Cations: Positive ions.

Collision theory: The theory that moving particles react when they collide with sufficient energy and in the correct orientation.

Compound: A substance made up of two or more different atoms (or ions) joined together by bonds.

Condensation polymerisation: Polymerisation occurring when monomers combine together with the elimination of small molecule.

Condensing: The change of state from gas to liquid.

Conductors (electrical): Substances that have a low resistance to the passage of electricity.

Corrosion: The gradual reaction and 'eating away' of a metal inwards from its surface caused by another substance.

Covalent bond: A shared pair of electrons between two atoms.

Cracking: The decomposition of larger alkane molecules into a mixture of smaller alkanes and alkenes.

Delocalised electrons: Electrons that are not associated with any particular atom.

Diatomic: Molecules containing two atoms.

Diffusion: The spreading movement of one substance through another due to the random movement of the particles.

Displayed formula: Shows how all the atoms and all bonds in a compound are arranged.

Dot-and-cross diagram: A diagram showing the electronic configuration of atoms, ions, or molecules.

Double bond: Two covalent bonds between the same two atoms.

Ductile: Can be drawn into wires.

Electrochemical series: The order of reactivity of metals, with the most reactive at the top.

Electrodes: Rods that conduct electric current to and from an electrolyte.

Electrolysis: The decomposition of a compound when molten or in solution by an electric current.

Electrolyte: A molten ionic compound or a solution containing ions that conducts electricity.

Electron: The negatively charged particles arranged in electron shells (energy levels) outside the nucleus of an atom.

Electron shells: Particular areas surrounding the nucleus that contain one or more electrons.

Electroplating: Coating of the surface of one metal with a layer of another, usually less reactive, metal using electrolysis.

Element: A substance made up of only one type of atom that cannot be broken down into anything simpler by chemical reactions.

Empirical formula: Shows the simplest whole number ratio of atoms or ions in a compound.

Endothermic reaction: A reaction that absorbs energy from the surroundings.

Energy profile diagram: Diagram showing the heat energy content of the reactants and products on the vertical axis and the reaction pathway on the horizontal axis.

Enthalpy change: The heat energy exchanged between a chemical reaction and its surroundings at constant pressure.

Enzymes: Biological catalysts.

Ester: A compound with the formula R–COO–R′ formed by the reaction of an alcohol with a carboxylic (alkanoic) acid.

Esterification: Making an ester by the reaction of an alcohol with a carboxylic acid.

Evaporation: The change of state from liquid to vapour that takes place below the boiling point of a liquid.

Exothermic reaction: A reaction that releases energy to the surroundings.

Fermentation: The breakdown of organic materials by microorganisms with effervescence and the release of heat energy.

Filtrate: The solution passing through a filter paper when a mixture of solid and solution are filtered.

Flue gas desulfurisation: Removal of sulfur dioxide arising from burning fossil fuels containing sulfur in industry.

Fraction: A product of petroleum distillation that is a mixture of hydrocarbons having a limited range of molar masses and boiling points.

Fractional distillation: A method used to separate two or more liquids with different boiling points from each other using a distillation column.

Freezing: The change of state from liquid to solid.

Functional group: A group that is characteristic of a given homologous series.

General formula: A formula that applies to all members of a given homologous series.

Giant molecular structure: A structure having a three-dimensional network of covalent bonds.

Global warming: The heating of the atmosphere caused by absorption and re-emission of infrared radiation by greenhouse gases.

Greenhouse gases: Gases that are good absorbers of infrared radiation and cause global warming.

Group: A vertical column in the Periodic Table.

Half-equations: Equations showing the oxidation and reduction reactions separately to show the loss or gain of electrons.

Halogens: The elements in Group VII.

Homologous series: A group of compounds with the same general formula and the same functional group.

Hydrocarbons: Compounds containing only carbon and hydrogen atoms.

Hydrogenation: A reaction involving the addition of hydrogen to a compound.

Hydrolysis: The breakdown of a compound by water, often catalysed by acids or alkalis.

Incomplete combustion: Combustion when air or oxygen is in limited supply.

Indicator: See acid–base indicator.

Insulators: Non-conductors.

Ion: A particle formed when an atom or group of atoms has lost or gained one or more electrons, making the particle positively or negatively charged.

Ionic bond: The strong force of attraction between oppositely charged ions.

Ionic equation: A symbol equation that shows only those ions that take part in a reaction.

Isotopes: Atoms of elements with the same number of protons but different numbers of neutrons.

Kinetic particle theory: The idea that particles are in constant motion.

Lustrous: Having a shiny surface.

Macromolecules: Very large molecules made up of repeating units.

Malleable: Can be shaped by hitting.

Mass number: The number of protons + the number of neutrons in an atom.

Melting: The change of state from solid to liquid.

Metallic bond: A bond formed by the attractive forces between the delocalised electrons and the positive ions in a metallic structure.

Metallic conduction: The movement of mobile electrons through the metal lattice when a voltage is applied.

Mixture: Two or more elements or compounds that are not chemically bonded together and can usually be separated by physical means.

Molar concentration: The number of moles of solute dissolved in a solvent to make 1 dm^3 of a solution.

Molar gas volume: The volume of a mole of gas at r.t.p. or s.t.p.

Molar mass: The mass of a substance in moles.

Mole: The amount of substance that has the same number of particles (atoms, ions, or electrons) as there are atoms in exactly 12 grams of the carbon-12 isotope.

Molecular equation: A full symbol equation.

Molecular formula: Shows the number of atoms of each particular element in one molecule of a compound.

Molecule: A particle containing two or more atoms bonded together. The atoms can be the same or different.

Monomers: The small molecules that react and bond together to form a polymer.

Neutralisation: The reaction between an acid and a base to form a salt and water.

Neutral oxide: An oxide that does not react with acids or alkalis.

Neutron: The neutral particle in the nucleus of an atom.

Noble gas configuration: A complete outer shell of electrons in an atom or ion so that the species has the electronic structure of one of the noble gases.

Nucleus: A tiny particle in the centre of an atom containing protons and neutrons.

Oxidation: The gain of oxygen or loss of electrons by a substance.

Oxidation number: A number given to each atom or ion in a compound to show the degree of oxidation.

Oxidising agent: A substance that accepts electrons and gets reduced during a chemical reaction.

Paper chromatography: A method used to separate a mixture of different dissolved substances depending on the solubility of the substances in the solvent and their attraction to paper.

Percentage yield:
$$\frac{\text{amount of required product obtained}}{\text{maximum amount of product expected}} \times 100$$

Periodic Table: Arrangement of elements in order of increasing atomic number so that most Groups contain elements with similar properties.

Periodicity: The regular occurrence of similar properties of the elements in the Periodic Table so that some Groups have similar properties or a trend in properties.

Period: A horizontal row in the Periodic Table.

Petroleum: A thick liquid mixture of unbranched, branched, and ring hydrocarbons extracted from beneath the Earth's surface.

Photochemical reaction: A reaction that depends on the presence of light.

pH scale: A scale of numbers from 0 to 14 used to describe how acidic or alkaline a solution is.

Physical properties: Properties that do not generally depend on the amount of substance present.

Pollution: Contaminating materials introduced into the natural environment (earth, air, or water).

Polyamide: Condensation polymer containing –NH–CO– linkages.

Polyester: Condensation polymer containing –COO– linkages.

Polymerisation: The conversion of monomers to polymers.

Polymers: Macromolecules made up by linking at least 50 monomers.

Precipitate: The solid obtained in a precipitation reaction.

Precipitation reaction: A reaction in which a solid is obtained when solutions of two soluble compounds are mixed.

Protons: The positively charged particles in the nucleus of an atom.

Radioactive isotopes: Isotopes with unstable nuclei, which break down.

Rate of reaction: The change in concentration of a reactant or product with time at a stated temperature.

Redox (reaction): A chemical reaction in which one reactant is oxidised and another is reduced.

Reducing agent: A substance that loses electrons and gets oxidised during a chemical reaction.

Reduction: The loss of oxygen or gain of electrons by a substance.

Relative atomic mass: The weighted average mass of naturally occurring atoms of an element on a scale on which an atom of carbon-12 has a mass of exactly 12 units.

Relative formula mass: The relative mass of one formula unit of a compound on a scale on which an atom of the carbon-12 isotope has a mass of exactly 12 units.

Glossary

Relative molecular mass: The relative mass of one molecule of a compound on a scale on which an atom of the carbon-12 isotope has a mass of exactly 12 units.

Residue: The solid remaining on the filter paper when a mixture of solid and solution are filtered.

r.t.p.: Room temperature and pressure (20 °C and 1 atmosphere pressure).

Rusting: Corrosion of iron and iron alloys caused by the presence of both water and oxygen.

Salt: A compound formed when the hydrogen in an acid is replaced by a metal or ammonium ion.

Saturated compounds: Organic compounds with only single carbon–carbon bonds.

Separating funnel: Piece of apparatus used to separate immiscible liquids that have different densities.

Simple distillation: The separation of a liquid from a solid, which involves the processes of boiling and condensation using a condenser.

Solubility: The number of grams of solute needed to form a saturated solution per 100 grams of solvent used.

Solute: A substance that is dissolved in a solvent.

Solution: A uniform mixture of two or more substances.

Solvent: A substance that dissolves a solute.

Sonorous: Rings when hit with a hard object.

Spectator ions: Ions that appear in a chemical equation but do not take part in the reaction.

Standard concentration: A concentration of 1 mole of substance in 1 dm^3 of solution under standard conditions.

State symbols: Letters put after a chemical formula showing whether it refers to a solid, liquid, gas, or aqueous solution.

Strong acid: An acid that ionises completely in solution.

Strong base: A base that ionises completely in solution.

Structural formula: Shows the way the atoms are arranged in a molecule with or without showing the bonds.

Structural isomers: Compounds with the same molecular formula but different structural formulae.

Sublimation: The direct conversion of a solid to a gas or gas to a solid without the liquid state being formed.

Substitution reaction: A reaction in which one atom or group of atoms replaces another.

Thermal decomposition: The breakdown of a compound when heated.

Titration: A method used to determine the amount of substance present in a given volume of solution (e.g. of acid or alkali).

Titre: The final burette reading minus the initial burette reading in a titration.

Triple bond: Three covalent bonds between the same two atoms.

Unbranched hydrocarbons: Hydrocarbons with carbon atoms linked in a chain without alkyl side groups.

Unsaturated compounds: Organic compounds containing double or triple carbon–carbon bonds (in addition to single bonds).

Volatile: Easily evaporated at room temperature.

Weak acid: An acid that only partially ionises in solution.

Weak base: A base that only partially ionises in solution.

Answers

Unit 1.1

LL In solids the particles are **regularly** arranged and close to each other. The particles only **vibrate**. They do not **move** from place to place. In liquids, the particles are not arranged in a **fixed** pattern and are **close** together. The particles move by **sliding** over each other. In gases, the particles are **far** apart and are able to move **everywhere** rapidly. (1 mark for each word in the correct place)

1. Box B: (solid): particles touching each other [1]
    particles arranged regularly / in more than 1 regular row [1]
    Box C: (liquid): particles touching each other [1]
    particles arranged irregularly / not in rows [1]
2. **a.** Diagram showing plunger pushed down so volume smaller **and** the same number of particles randomly arranged [1]
    b. Fewer particles hit the wall [1]
    Every second [1]
    So the force per unit area on the wall is less [1]
    c. Pressure decreases [1]

Unit 1.2

LL When a solid is heated, the increase in **energy** makes the particles **vibrate** more. The forces of **attraction** between the particles are weakened. At the **melting** point, these forces are **weak** enough for the particles to be able to move and slide over each other. When the liquid is at its **boiling** point, the particles have enough energy to **escape** from the **surface** of the liquid. (1 mark for each word in the correct place)

1. **a.** A: melting / fusion [1]
    B: boiling / evaporation [1]
    C: freezing [1]
    D: condensing [1]
    b. i. Methane [1]
    ii. Naphthalene [1]
    Melting point is above room temperature [1]
    iii. Ethanol [1]
    Melting point is below room temperature **and** boiling point is above room temperature / room temperature is between melting point and boiling point [1]
    c. A: solid **and** liquid [1]
    B: liquid [1]
    C: liquid **and** vapour / liquid **and** gas [1]
    D: vapour / gas [1]

Unit 1.3

LL Brownian; Gases; Random; Kinetic; Collide; Move; Particle; Diffuse (1 mark each)

1. **a.** Gases [1]
    b. Gases and liquids [1]
    c. Solids [1]
    d. Gases [1]
2. **a.**

    from here   to here

    Random movement shown, e.g. particle moves in 4 or more different directions [1]
    All lines are straight [1] (arrows need not be shown)
    b. Pollen grains in water / clay particles in water / other suitable example [1]
    c. More particles in the air bombard (hit) the dust particle on one side than on another (or hit with greater force) [1]
    Dust particles move in direction of the greater number of hits [1]
    Particles in the air move randomly so the direction of the movement of the dust particles is also random [1]
3. Chlorine [1] It has the highest relative molecular mass [1]

Unit 1.4

LL In liquids and gases, the **particles** are constantly moving and **changing** direction when they **hit** other particles. We say that they move **randomly**. Diffusion is the random **movement** of particles in any direction so that they get **mixed** up. Diffusion in **gases** is faster than in **liquids** because the particles move faster in gases. (1 mark for each word in the correct place)

1. **a. i.** Dissolving [1]
    ii. Diffusion [1]
    b. Particles (of dye and water) move randomly / move in any direction [1]
    Dye particles spread out [1]
    Overall movement of the dye is from area of high concentration (of the dye particles) to lower concentration (of dye particles) [1]
2. White solid forms where hydrogen chloride reacts with ammonia [1] Hydrogen chloride has a higher relative molecular mass than ammonia. [1] So rate of diffusion of hydrogen chloride less than that of ammonia. [1]

Unit 1.5

LL Test tube; Burette; Pipette; Flask; Beaker; Syringe; Timer; Balance (1 mark each)

1. **a.** A burette [1] B volumetric flask [1] C measuring cyclinder [1]
    D volumetric pipette [1]
    b. i. B [1]
    ii. D (or A) [1]
2. B [1]

Unit 1.6

LL The method of separating a **mixture** of coloured substances using **filter** paper is called chromatography. The colours **separate** if they have different **solubilities** in the solvent and different degrees of **attraction** for the filter paper. Chromatography can also be used to separate colourless substances. These are shown up after chromatography by **spraying** the paper with a **locating** agent. (1 mark for each word in the correct place)

1.

Chromatography paper
Solvent

    Chromatography paper dipping in solvent [1]
    Chromatography paper labelled [1]
    Solvent labelled [1]
2. **a.** So that the ink doesn't spread up the paper / graphite/pencil 'lead' doesn't dissolve in solvent [1]
    b. 3 [1]
    c. Ser and Gly [1]
    d. $\dfrac{\text{distance from centre of spot to baseline}}{\text{distance from solvent front to baseline}} = 0.6$ [1]
    e. About half way between Cys and Ser/Gly [1]

Unit 1.7

LL The melting and **boiling** points of **pure** substances are sharp. They melt and boil at **exact** temperatures. The melting and boiling points of **impure** substances are not sharp. They melt over a **range** of temperatures. The boiling point of a liquid is **increased** if impurities are present. The melting point of a liquid is **decreased** if impurities are present. (1 mark for each word in the correct place)

1. **a.** Oxygen gas [1] sodium chloride crystals [1]
    b. Some impurities may be harmful [1]
    c. Any values from −15 °C to −1 °C
2. **a.** Pure sulfur: solidifies at 119 °C
    Pure sulfur: has a sharp boiling point
    Impure sulfur: melts over 4 °C temperature range
    Impure sulfur: turns to a vapour at 450 °C
    (2 marks if all 4 correct; 1 mark if 2 correct)
    b. i. In solder the tin is impure / the lead is impure [1]
    Impurities lower the melting point [1]
    ii. Less energy is used in melting the solder (than using tin or lead alone) [1]

159

Unit 1.8

LL A with 3, B with 4, C with 5, D with 1, E with 2 (3 marks if all 5 correct, 2 marks if 3 or 4 correct, 1 mark if 1 or 2 correct)

1. a. i. In order downwards:
      Filter paper [1]
      Filter funnel [1]
      Flask [1]
   ii. Residue on filter paper [1]
      Filtrate is liquid in flask [1]
2. a. BGFEADC (2 marks) (1 mark if 1 pair reversed)
   b. Too much water may dissolve the crystals. [1]

Unit 1.9

LL There is a range of **temperatures** in the distillation column, **lower** at the top and **higher** at the bottom. When **vaporised** the more **volatile** alcohols move **further** up the column than the less volatile alcohols. When the alcohol reaches the **condenser** it changes from vapour to **liquid**. The alcohols are collected one by one in the **receiver**, those with the lower **boiling** points condensing before those with higher ones.
(1 mark for each word in the correct place)

1. a. i. Distillation flask on left [1]
      Distillate in beaker on right [1]
      Slanting tube labelled as condenser [1]
      Cold water enters the bottom of the condenser [1]
   ii. Arrow under the gauze [1]
   b. i. Salt and water have very different boiling points / salt has a high boiling point and water has a low boiling point [1]
   ii. The vapours would condense together / at the same time [1]
2. a. filtration [1]
   b. simple distillation [1]
   c. fractional distillation [1]

Unit 2.1

LL Atoms are the **smallest** particles of matter that can take part in a **chemical** change. Each atom consists of a **nucleus** made up of protons and **neutrons**. Outside the nucleus are the **electrons**. These are **arranged** in electron **shells** or energy **levels**.
(1 mark for each word in the correct place)

1. No nucleus in Thomson's model [1]
   Positive charge spread out rather than being in the nucleus [1]
   Electrons not in shells [1]
2. a. Most of the alpha particles went straight through the foil [1]
   b. A few alpha particles changed course [1]
      The positive charge on the nucleus repels the positive charge on the alpha particle / like charges repel [1]

Unit 2.2

LL A with 4, B with 1, C with 2, D with 3 (all 4 correct 2 marks, 2 or 3 correct 1 mark)

1. a. Isotopes are **atoms** of the same **element** with the same number of **protons** but different numbers of **neutrons**. (1 mark each correct word)
   b. i. 1 [1]
   ii. It has no neutrons [1]
2. a. As the time increases equally / 14 days the mass decreases by a half [2]
      (BUT: 1 mark for: As time increases mass decreases)
   b. 25 cpm [1]
3. Any 1 use (1 mark) e.g. smoke detectors / finding leaks in pipes / measuring thickness of paper

Unit 2.3

LL The **arrangement** of the electrons in shells is called the electron arrangement or electron **distribution**. An atom of fluorine has nine electrons, **two** in the first shell and **seven** in the second shell. Atoms of elements in the same **Group** have the same number of electrons in their **outer** shell. As we move across a **Period**, each atom has **one** more **electron** in its outer shell than the element before it.
(1 mark for each word in the correct place)

1.

element	number of electrons in an atom	electron distribution
nitrogen	7	2,5
oxygen	8	2,6
fluorine	9	2,7
neon	10	2,8
sodium	11	2,8,1
argon	18	2,8,8
calcium	20	2,8,8,2

(1 mark for correct number of electrons in column 1. 1 mark for each correct electron distribution)

2.

2,8,3 aluminium	2,4 carbon	2,8,7 chlorine	2 helium
2,8,2 magnesium	2,8 neon	2,8,5 phosphorus	2,8,8,1 potassium

(1 mark each correct structure. Electrons should be drawn in shells and paired.)

Unit 2.4

LL An element ..... is a substance containing only ..... one type of atom ..... which cannot be broken down further .....
by chemical means. [1]
A compound ..... is a substance containing two or more ..... types of atom ..... which are chemically ..... combined (bonded). [1]

1.

compound	mixture
The **elements** cannot be **separated** by **physical** means.	The substances in it can be **separated** by **physical** means.
The properties are **different** from those of the **elements** that went to make it.	The properties are the **average** of the substances in it.
The elements are **combined** in a **definite** proportion by mass.	The substances can be **present** in **any** proportion by mass.

(12 correct = 6, 10 or 11 correct = 5, 8 or 9 correct = 4, 6 or 7 correct = 3, 4 or 5 correct = 2, 2 or 3 correct = 1)

2. A compound [1] B element [1] C element [1]
   D mixture [1] E compound [1] F mixture [1]

Unit 2.5

LL Malleable; Ductile; Conduct; Lustre; Sonorous; Dense (1 mark each)

1. a. conducts electricity [1] ductile [1] malleable [1] shiny [1]
   b. i. Does not conduct electricity / shatters when hit [1]
   ii. Conducts heat / very high melting point [1]
   iii. Low melting point [1]
2. Aluminium [1] has the lowest density [1]

Unit 3.1

LL A sodium chloride **lattice** is a **regular** arrangement of **positive** sodium ions and **negative** chloride ions that **alternate** with each other. The ions are held together by **strong** ionic **bonds**. This structure is called a **giant** ionic structure.
(1 mark for each word in the correct place)

1. 2,8 2,8,8 2 2,8,18,8
(1 mark each)

Answers

2.

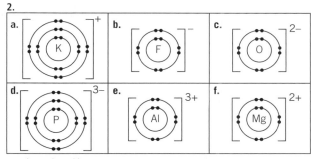

(1 mark each)

3. 8 electrons in outer shell of Mg [1]
8 electrons in outer shell of S [1]
2– charge on S [1]

Unit 3.2

LL A with 3, B with 1, C with 4, D with 2 (2 marks if 4 correct, 1 mark if 2 or 3 correct)

1. CO   Cl$_2$ N$_2$ O$_2$ [2]
(1 mark if 3 correct)

2. a.

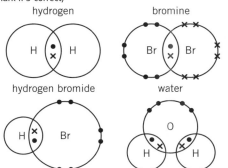

hydrogen   bromine

hydrogen bromide   water

(1 mark each)

b. i. Two around the hydrogen and eight around the others [1]
(you will not get this mark if you just write eight)
ii. The electron shells are complete / full [1]
This is a stable structure / electrons cannot easily be lost or gained [1]

Unit 3.3

LL A covalent bond is formed when **two** atoms combine. It forms because of the **strong** force of **attraction** between the **nucleus** of one atom and the outer **electrons** of the atom next to it.
(1 mark for each word in the correct place)

1.

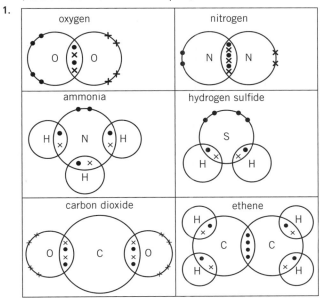

oxygen   nitrogen

ammonia   hydrogen sulfide

carbon dioxide   ethene

(1 mark each)

Unit 3.4

LL Simple covalent compounds have **low** melting points because the **intermolecular** attractive forces are **weak**. Ionic compounds have **high** melting points because of the **strong** forces of **attraction** between the positive and negative **ions**.
(1 mark for each word in the correct place)

1. Barium oxide: ionic
Carbon tetrachloride: covalent
Potassium bromide; ionic
Carbon disulfide: covalent
Octane: covalent
(2 marks if all correct and 1 mark if one error)

2. A with 4, B with 7, C with 1, D with 2, E with 6, F with 3, G with 5
(all 7 correct = 3 marks, 5 or 6 correct = 2 marks, 3 or 4 correct = 1 mark)

Unit 3.5

LL Allotrope; Layers; Tetrahedral; Giant; Delocalise; Covalent (1 mark each)

1. a. Diamond: B, C, D [1]
Graphite: A, D, F [1]
Silicon dioxide: B, C, D [1]
b. Diamond and graphite: Allotropes are different forms of the same element, in this case carbon. [1]

2. A with 4, B with 1, C with 5, D with 2, E with 3
(all correct = 2 marks but 3 or 4 correct = 1 mark)

3. Diamond: tetrahedral [1] Graphite: hexagonally arranged / arranged in rings of 6 [1]

Unit 3.6

LL Atoms of metallic elements are generally arranged in closely packed **layers**. The outer **electrons** tend to **move** away from their atoms to form a 'sea' of **delocalised** electrons around the **positive** ions. When a **voltage** is applied **across** the metal the delocalised electrons are able to move.
(1 mark for each word in the correct place)

1. a.

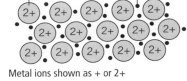

metal ion   free electron

Metal ions shown as + or 2+ [1]
Metal ions labelled [1]
Electrons shown as dots randomly dispersed between the metal ions [1]
Electrons labelled free electrons / delocalised electrons / mobile electrons [1]
b. When force applied the force of attraction between metal ions and mobile electrons is overcome [1]
The layers slide over each other [1]
(When force removed) forces of attraction between metal ions and electrons hold layers in new position [1]

2. a. D [1]
b. A, B, and C [1] Idea that metals are to the left of Group IV / to the left of giant covalent structures (in the Periodic Table) [1]
c. E and F [1] They have low melting points / they are to the right of Period IV [1]

Unit 4.1

LL (1 mark each for any 6) selenium; sulfur; nickel; carbon; iodine; argon; nitrogen

1. **a.**

								H 1	
Li 1					B 3	C 4	N 3	O 2	F 1	Ne 0
Na 1	Mg 2				Al 3		S 2	Cl 1	
K 1	Ca 2		transition elements variable	Zn 2				Br 1	

(1 mark for each column correct (= 8), 1 mark for H, 1 mark for transition elements + Zn)

  b. Li, Na, K, Mg, Ca, Zn, Al, and H [1]
  c. N, O, S, F, Cl, Br (H) [1]
  d. **i.** H_2S [1]
    ii. B_2O_3 [1]
    iii. CS_2 [1]
    iv. CBr_4 [1]
    v. Ca_3N_2 [1]
    vi. Al_2O_3 [1]

2. R magnesium iodide [1] S Strontium hydroxide [1]
  T iron(II) sulfate [1] U zinc nitrate [1]
  V ammonium carbonate [1] W calcium hydrogencarbonate [1]

Unit 4.2

LL Lead; Carbon; Bromine; Tin; Neon; Calcium; Iron; Nickel; Uranium; Sulfur (1 mark each)

1. **a.** Gain [1]
  b. Loss of electrons [1]
2. **a.** A $MgBr_2$ [1]   B Na_2O [1]   C HCl [1]   D $AlCl_3$ [1]
    E K_3N [1]   F CaS [1]   G Al_2S_3 [1]   H Fe_2O_3
  b. J $Mg(NO_3)_2$ [1]   K K_2SO_4 [1]   L NH_4NO_3 [1]
    M $(NH_4)_2SO_4$ [1]   N $Ca(OH)_2$ [1]   O $NaHCO_3$ [1]
    P $Al(NO_3)_3$ [1]   Q Li_2CO_3 [1]

Unit 4.3

LL A with 3; B with 1; C with 2; D with 4 (2 marks if all correct, 1 mark if 2 or 3 correct)

1. **a.** $O_2 + 2H_2 \rightarrow 2H_2O$
    Correct balance [1] correct use of + and $\rightarrow$ [1]
  b. $2C + O_2 \rightarrow 2CO$
    Correct balance [1] correct use of + and $\rightarrow$ [1]
2. **a.** $2K + Br_2 \rightarrow 2KBr$ [1]
  b. $4Al + 3O_2 \rightarrow 2Al_2O_3$ [1]
  c. $4Na + O_2 \rightarrow 2Na_2O$ [1]
  d. $N_2 + 3H_2 \rightarrow 2NH_3$ [1]
  e. $2Rb + 2H_2O \rightarrow 2RbOH + H_2$ [1]

Unit 4.4

LL A with 4; B with 3; C with 2; D with 1 (2 marks if all correct, 1 mark if 2 or 3 correct)

1. **a.** Na^+ and OH^- [1]   **b.** Mg^{2+} and Cl^- [1]   **c.** Ba^{2+} and NO_3^- [1]
  d. Cu^{2+} and SO_4^{2-} [1]   **e.** Al^{3+} and O^{2-} [1]   **f.** Fe^{2+} and OH^- [1]
2. **a.** Ions $Cu^{2+} + 2Cl^-$ [1] $(2Na^+) + 2Cl^-$ [1]
    Cancel: $Cu^{2+} + \cancel{2Cl^-}$ and $\cancel{(2Na^+)} + \cancel{2Cl^-}$ [1]
    Equation: $Cu^{2+}(aq) + 2OH^-(aq) \rightarrow Cu(OH)_2(s)$ [1]
  b. Ions: $Ba^{2+} + 2Cl^-$ $Mg^{2+} + SO_4^{2-}$ [1] $\rightarrow Mg^{2+} + 2Cl^-$ [1]
    Cancel: $Ba^{2+} + \cancel{2Cl^-}$ $\cancel{Mg^{2+}} + SO_4^{2-}$ [1] $\rightarrow \cancel{Mg^{2+}} + \cancel{2Cl^-}$ [1]
    Equation $Ba^{2+}(aq) + SO_4^{2-}(aq) \rightarrow BaSO_4(s)$ [1]
  c. Ions: $2K^+ + 2I^-$ [1] $\rightarrow 2K^+ + 2Cl^-$ [1]
    Equation: $Cl_2(aq) + 2I^-(aq) \rightarrow I_2(aq) + 2Cl^-(aq)$ [1]

Unit 5.1

LL Relative atomic mass (symbol A_r) is the **average** mass of naturally occurring **atoms** of an element on a scale on which the **carbon-12** atom has a mass of exactly **twelve** units.
(1 mark for each word in the correct place)

1. **a.** (2 marks for each correct M_r. If 2 not scored, 1 mark for correct number of atoms)

compound	number of each atom	A_r of atom	M_r calculation
phosphorus trichloride PCl_3	P = 1 Cl = 3	P = 31 Cl = 35.5	$M_r = \dfrac{1 \times 31}{137.5} \dfrac{3 \times \mathbf{35.5}}{}$
magnesium hydroxide $Mg(OH)_2$	Mg = 1 O = 2 H = 2	Mg = 24 O = 16 H = 1	$M_r = \dfrac{1 \times 24}{58} \dfrac{2 \times 16}{} \dfrac{2 \times 1}{}$
ethanol C_2H_5OH	C = 2 H = 1 O = 16	C = 12 H = 1 O = 16	$M_r = \dfrac{2 \times 12}{46} \dfrac{6 \times 1}{} \dfrac{1 \times 16}{}$
ammonium sulfate $(NH_4)_2SO_4$	N = 2 H = 8 S = 1 O = 4	N = 14 H = 1 S = 32 O = 16	$M_r = \dfrac{2 \times 14}{132} \dfrac{8 \times 1}{} \dfrac{1 \times 32}{} \dfrac{4 \times 16}{}$
glucose $C_6H_{12}O_6$	C = 6 H = 12 O = 6	C = 12 H = 1 O = 16	$M_r = \dfrac{6 \times 12}{180} \dfrac{12 \times 1}{} \dfrac{6 \times 16}{}$

  b. **i.** 342 [1]   **ii.** 183 [1]   **iii.** 220 [1]

Unit 5.2

LL A with 2, B with 3, C with 4, D with 1

1. $\dfrac{12}{48} \times 80 = 20$ g
  (1 mark)   (1 mark)
  b. 4.8 g [1]
  c. 280 g [1]
2. (1 mark for each correct answer)

element or compound	formula mass, M_r	mass taken / g	number of moles
O_2	32	4	0.125 [1]
NaCl	58.5 [1]	11.7	0.2 [1]
$CaSO_4$	136 [1]	27.2	0.2 [1]
P_2O_5	142 [1]	56.8 [1]	0.4
CO_2	44 [1]	4.4 [1]	0.1
P_4	124 [1]	86.8	0.7 [1]
CH_4	16 [1]	384 [1]	24.0

Unit 5.3

LL To find the **number** of moles of a compound, we need to know the mass of compound taken and the **relative** molecular mass of the compound. The relative molecular mass is found by **adding** together the relative **atomic** masses of all the **atoms** (or ions) in the compound. The number of **moles** is found by **dividing** the mass of compound taken by the relative molecular mass.
(1 mark for each word in the correct place)

1. $4 \times 34 \rightarrow (4 \times 31) + (6 \times 2)$ [1]
  $\mathbf{136}$ g $\rightarrow \mathbf{124}$ g + $\mathbf{12}$ g [1]
2. **a.** 5 [1]
  b. 1 [1]
  c. 2 [1]
  d. mol $I_2O_5 = 20.04/334 = 0.06$ [1] So $5 \times 0.06 = 0.30$ mol CO_2 [1]
    mass $= 0.30 \times 44 = 13.2$ g [1]
  e. mol $CO_2 = 21/28 = 0.75$ [1] So $0.75 / 5 = 0.15$ mol I_2 [1]
    mass $= 0.15 \times 254 = 38.1$ g [1]
2. **a.** $232 - 168 = 64$ g [1]
  b. $64/16 = 4$ mol O [1]
  c. $168/56 = 3$ mol [1]
  d. 3Fe:4O [1] Fe_3O_4 [1]

Unit 5.4

LL At room **temperature** and pressure, one **mole** of any gas occupies 24 **dm³** (24 000 cm³). Because there is always the same **number** of moles, there is also the same number of **molecules** in a given volume. So, 50 cm³ of chlorine and 50 cm³ of **oxygen** under the same **conditions** contain the same number of molecules.
(1 mark for each word in the correct place)

1. a. $\dfrac{2 \times 12}{(2 \times 12) + (6 \times 1)} \times 100 = \mathbf{80\%}$
   (2 marks for correct answer, 1 mark if answer wrong but working correct)

   b. $\dfrac{14}{17} \times 100 = 82.4\%$ (2 marks if correct, 1 mark if answer wrong but working correct)

   c. $\dfrac{40}{100} \times 100 = 40\%$ (2 marks if correct, 1 mark if answer wrong but working correct)

2. (1 mark for each correct answer)

gas	M_r of gas	mass of gas / g	moles of gas / mol	volume of gas / dm³
ammonia	17	8.5	**0.5** [1]	**12** [1]
oxygen	32	**640** [1]	**20** [1]	480
carbon dioxide	44	3.08	**0.07** [1]	**1.68** [1]
hydrogen chloride	**36.5** [1]	292	8	**192** [1]
ethane	30	**3.75** [1]	**0.125** [1]	3

Unit 5.5

LL **percentage** purity = $\dfrac{\mathbf{mass}\ \text{of}\ \mathbf{pure}\ \text{product}}{\text{mass of}\ \mathbf{impure}\ \text{product}} \times 100$

(1 mark for each word in the correct place)

1. a. 100 g/mol [1]
   b. 3.840 dm³ [1]
   c. 3.840/24 = 0.16 mol [1]
   d. 0.16 mol [1]
   e. 0.16 × 100 = 16 g [1]
   f. 16/18 (× 100) = 89% [1]
2. a. 122 g/mol [1]
   b. 24.4/122 = 0.2 mol [1]
   c. 0.2 mol [1]
   d. 136 g/mol [1]
   e. 0.2 × 136 = 27.2 g [1]
   f. 25.84/27.2 (× 100) = 95% [1]

Unit 5.6

LL The molecular formula of a **compound** shows the number of **each** type of atom in one **molecule**. The empirical formula shows the **simplest** ratio of **atoms** that **combine**.
(1 mark for each word in the correct place)

1. moles of Pb = 0.1 mol   moles of Cl = 0.4 mol [1]
   Divide by   Pb **0.1**   Cl **0.4** [1]
   lowest number   0.1   0.1
   of moles
   Result of division = 1   = 4
   Simplest ratio **1Pb:4Cl** [1] So empirical formula is **PbCl₄** [1]
2. a. HO [1]   b. Sb₂O₃ [1]   c. C₂H₅ [1]
3. A empirical formula mass 110 [1]   molecular formula P₄O₆ [1]
   B empirical formula mass 67.5 [1]   molecular formula S₂Cl₂ [1]
   C empirical formula mass 30 [1]   molecular formula C₂H₄O₂ [1]

Unit 5.7

LL concentration in **moles** per dm³ = $\dfrac{\mathbf{amount}\ \text{of}\ \mathbf{solute}\ \text{in moles}}{\text{volume in}\ \mathbf{dm^3}}$

1.

solute	M_r of solute	mass of solute / g	volume of solution cm³ or dm³	concentration of solution mol/dm³
sodium hydroxide	40	8	250 cm³	**0.8** [1]
silver nitrate	170	**17** [1]	200 cm³	0.5
copper(II) sulfate	160	40	**2.0 dm³** [1]	0.125

2. a. moles of acid = $\underline{\mathbf{0.10}} \times \dfrac{\mathbf{12.2}}{1000} = \underline{\mathbf{1.22 \times 10^{-3}}}$ mol H₂SO₄ [1]
   b. i. 2 [1]
      ii. $1.22 \times 10^{-3} \times 2 = 2.44 \times 10^{-3}$ mol NaOH [1]
   c. $\dfrac{2.44 \times 10^{-3}}{0.025} = 0.098$ mol/dm³
      (2 marks for correct answer, 1 mark for 0.025 if correct answer not obtained)

Unit 6.1

LL One mark for each correct word

	D	E	C	O	M	P	O	S	E
C				A					D
E	L	E	C	T	R	O	D	E	O
L				H					N
L			I	O	N				A
			D						
B	A	T	T	E	R	Y			

1. A battery / cell(s) / power supply [1]
   B anode [1]
   C cathode [1]
   D electrolyte [1]
2. a. Electrolysis: breakdown of ionic substance when molten or in solution [1] by passage of electricity [1]
   b. Electrolyte: liquid which conducts electricity [1]
3. (1 mark each 'cell' correct)

electrolyte	cathode (−) product	anode (+) product	observations at the anode
molten zinc bromide	zinc	bromine	red-brown fumes / solution
molten magnesium chloride	magnesium	chlorine	yellow-green / green fumes
molten calcium oxide	calcium	oxygen	bubbles (colourless)
molten lead iodide	lead	iodine	brown solution / purple vapour

Unit 6.2

LL If a metal is more reactive than hydrogen, the metal ions stay in solution and hydrogen, arising from hydrogen ions in water, bubbles off.
(2 marks if all correct, 1 mark if one pair of phrases in the incorrect place)

1. (1 mark for each 'cell' correct)

electrolyte	cathode (−) product	anode (+) product	observations at the anode
concentrated KCl(aq)	hydrogen	chlorine	bubbles of gas, green when collected
dilute H₂SO₄(aq)	hydrogen	oxygen	colourless bubbles
very dilute NaCl(aq)	hydrogen	oxygen	colourless bubbles
concentrated HCl(aq)	hydrogen	chlorine	bubbles of gas, green when collected

2. a. B chlorine [1] C hydrogen [1]
   b. A [1]
   c. Chlorine discharged more readily [1] than oxygen (from hydroxide ions in water) [1]

Unit 6.3

LL During electrolysis, positive ions move towards the **cathode** where they gain **electrons**. This is a **reduction** reaction. Negative ions **move** to the **anode** where they **lose** electrons. This is an **oxidation** reaction.
(1 mark for each word in the correct place)

1.

a. molten zinc bromide

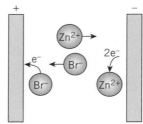

b. dilute sulfuric acid

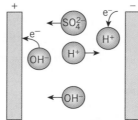

c. concentrated hydrochloric acid

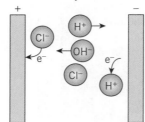

d. aqueous copper(II) sulfate

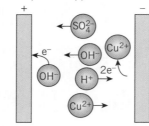

a. Zinc ions moving to cathode and bromide ions to anode [1]
Bromide ions donating electrons to anode and zinc ions taking electrons from cathode [1]

b. Hydrogen ions moving to cathode and hydroxide and sulfate ions to anode [1]
Hydroxide ions donating electrons to anode and hydrogen ions taking electrons from cathode [1]

c. Hydrogen ions moving to cathode and chloride and hydroxide ions to anode [1]
Chloride ions donating electrons to anode and hydrogen ions taking electrons from cathode [1]

d. Hydrogen ions and copper ions moving to cathode and hydroxide and sulfate ions to anode [1]
Hydroxide ions donating electrons to anode and copper ions taking electrons from cathode [1]

2. **a.** $Zn^{2+} + \underline{2e^-} \rightarrow Zn$ [1]

b. $\underline{2}Cl^- \rightarrow \underline{Cl_2} + 2e^-$ (1 mark for 2 and Cl_2, 1 mark for balance with electrons)

c. $\underline{2}H^+ + \underline{2e^-} \rightarrow \underline{H_2}$ (1 mark for 2 and H_2, 1 mark for balance with electrons)

d. $Al^{3+} + \underline{3e^-} \rightarrow \underline{Al}$ [1]

Unit 6.4

LL The electrolysis cell has an **impure** strip of copper as the **anode** and a pure strip of **copper** as the cathode. The **electrolyte** is a solution of copper(II) sulfate. At the anode, copper **atoms** lose electrons and go into **solution** as copper(II) **ions**. At the cathode, copper(II) ions **gain** electrons and are deposited on the **cathode** as copper atoms.
(1 mark for each word in the correct place)

1. **a.** Anode: $Cu \rightarrow Cu^{2+} + 2e^-$ (1 mark for correct species, 1 mark for balance)
Cathode: $Cu^{2+} + 2e^- \rightarrow Cu$ (1 mark for correct species, 1 mark for balance)

b. Cu^{2+} ions removed from cathode and Cu^{2+} formed at the anode [1]
At same rate [1]

3.

mass of the electrodes	anode: no change	[1]	anode: decreases	[1]
	cathode: increases slightly	[1]	cathode: large increase	[1]
appearance	anode: none / bubbles given off	[1]	anode: gets thinner	[1]
	cathode: goes pink / brown	[1]	cathode: gets thicker with lighter colour pink deposit	[1]
electrolyte	gets a lighter blue/ fades	[1]	remains the same depth of colour	[1]

Unit 6.5

LL The object to be **electroplated** is connected to the **negative** pole of the **power** supply. The object becomes the **cathode**. A strip of the plating **metal** is connected to the positive **pole** of the power supply. The **plating** metal is the anode. The **electrolyte** is a solution containing **ions** of the plating metal.
(1 mark for each word in the correct place)

1. **a.** A = rod, C = jug, E = liquid in which the anode and cathode dip (3 correct = 2 marks, 1 or 2 correct = 1 mark)

b. gains mass (slightly) [1] becomes silvery [1]

c. Anode: $Ag \rightarrow Ag^+ + e^-$ [1]
Cathode: $Ag^+ + e^- \rightarrow Ag$ [1]

2. **a.** Tin forms a layer over the iron / tin covers the iron [1]
Prevents water or oxygen from getting to the surface of the iron [1]

b. Makes them look attractive [1]

Unit 6.6

LL Bauxite; Oxygen; Graphite; Alumina; Cryolite; Carbon (1 mark each)

1. E = liquid in the cell into which rods are dipping [1]
C = layer on the inside next to liquid [1]
A = rods dipping into the liquid [1]
M = layer at the bottom of the cell [1]

2. Aluminium oxide has a very high melting point [1]
Cryolite dissolves the aluminium oxide [1]
Melting point of the aluminium oxide is lowered / less energy needed [1]

3. **a.** **i.** $Al^{3+} + 3e^- \rightarrow Al$ (1 mark for correct formulae, 1 mark for balance)

ii. $2O^{2-} \rightarrow O_2 + 4e^-$ (1 mark for correct formulae, 1 mark for balance)

b. $2Al_2O_3 \rightarrow 4Al + 3O_2$ (1 mark for correct formulae, 1 mark for balance)

Unit 6.7

LL Ceramics are useful insulators, not only because they resist the flow of electricity, but also because they have very high melting points.
(2 marks if all correct, 1 mark if one pair of phrases in the incorrect place)

1. **a.** **i.** (indicator) bulb / lamp labelled [1] cells / battery / power source labelled [1]

ii. arrows clockwise around circuit [1]

2. **a.** Good electrical conductor [1] low density / lightweight [1]

b. (Fairly good) electrical conductor [1] strong [1]

3. A with 3   B with 1   C with 4   D with 2 (2 marks if 4 correct, 1 mark if 2 or 3 correct)

Unit 7.1

LL A with 4; B with 3; C with 1; D with 2 (2 marks if all correct, 1 mark if 2 or 3 correct)

1. Separating iron from sulfur [1], melting zinc [1] Distilling plant oils [1]

2. **a.** endothermic [1]   **b.** exothermic [1]   **c.** endothermic [1]

d. exothermic [1]

3. **a.** Energy on the vertical axis of both L and M [1]
Reactants on lines on the left of both L and M [1]
Products on the lines on the right of both L and M [1]
Downward arrow between the two lines in L [1]
Upward arrow between the two lines in M [1]

b. The energy of the reactants is greater than the energy of the products [1]
So energy is released [1]

Unit 7.2

LL In an endothermic reaction, the energy **absorbed** in bond breaking is **greater** than the energy **released** when new **bonds** are formed. In an exothermic reaction, the energy given **out** when **new** bonds are formed is greater than the energy taken in when the bonds in the **reactants** are broken.
(1 mark for each word in the correct place)

1.

bonds broken (endothermic +) / kJ/mol		bonds formed (exothermic −) / kJ/mol	
$4 \times (C-H) = 4 \times 413 = $ **1652**	[1]	$2 \times (C=O) = 2 \times 805 = $ 1610	[1]
$\underline{2} \times (O=O) = \underline{2 \times 498} = $ **996**	[1]	$4 \times (O-H) = 4 \times 464 = $ 1856	[1]
Total	+ **2648**	Total =	−3466

Overall energy change = (+2648) + (−3466) = −818 kJ/mol [1]

2. Graph with two axes with energy labelled in kJ/mol (or J/mol) on vertical axis [1] $CH_4(g) + 2O_2(g)$ on horizontal line on left of diagram and $CO_2(g) + 2H_2O(l)$ on horizontal line on right of diagram, and product energy level below reactant energy level [1]
Arrow drawn downwards from horizontal line on left to horizontal line on right [1]

Unit 7.3

LL A with 4; B with 1; C with 2; D with 3 (2 marks if all correct, 1 mark if 2 or 3 correct)

1. a. C [1]
   b. B [1]
   c. i. B [1] it has the lowest density / it weighs least [1]
      ii. Hazard of the fuel e.g. how combustible it is / how poisonous it is / its state / whether it is solid, liquid, or gas [1]
2. a. $2H_2 + O_2 \rightarrow 2H_2O$ (1 mark for correct symbols, 1 mark for balance)
   b. $2C_2H_6 + 7O_2 \rightarrow 4CO_2 + 6H_2O$ (1 mark for correct symbols, 1 mark for balance)
   c. $C_7H_{16} + 11O_2 \rightarrow 7CO_2 + 8H_2O$ (1 mark for correct symbols, 1 mark for balance)
   d. $2H_2S + 3O_2 \rightarrow 2H_2O + 2SO_2$ (1 mark for correct symbols, 1 mark for balance)

Unit 7.4

LL Zinc is more reactive than copper, so zinc is better at releasing electrons, forming zinc ions and becoming the negative pole of the cell.
(2 marks if all correct, 1 mark if one pair of phrases in the incorrect place)

1. a. i. magnesium [1] The voltage difference between magnesium and copper is higher than that between tin and copper. [1]
      ii. 2.26 V [1]
   b. i. lithium [1]
      ii. silver [1]
      iii. zinc / iron [1]

Unit 7.5

LL A fuel cell consists of two **porous** electrodes coated with **platinum**. The electrolyte is either an acid or an **alkali**. Hydrogen and **oxygen** are bubbled through the porous electrodes where the **reactions** take place. When connected to an **external** circuit, **electrons** flow from the **negative** electrode to the positive electrode.
(1 mark for each word in the correct place)

1. a. Voltmeter [1]
   b. D [1]
   c. A [1]
   d. Arrow drawn on external circuit in direction from negative to positive electrode (left to right) [1]
2. a. $2H_2 + 4OH^- \rightarrow 4H_2O + 4e^-$ (1 mark for correct symbols, 1 mark for balance)
   b. $O_2 + 2H_2O + 4e^- \rightarrow 4OH^-$ (1 mark for correct symbols, 1 mark for balance)
3. (1 mark each for any two) no pollutants formed or only water formed / lighter in weight / more efficient or fewer moving parts / produce more energy per g of fuel burnt

Unit 8.1

LL To find the rate of reaction we can either measure how **quickly** the reactants are **used** up or how quickly the **products** are formed. To calculate the **rate** of reaction we need to find out how some measurement changes with **time**. For example, the **volume** of gas given off per **second**, or how the mass of the reaction mixture **decreases** with time.
(1 mark for each word in the correct place)

1. B, C, D, A [1]
2. a. Copper reducing in size [1]
      Solution getting darker [1]
      Gas given off [1]
   b. (1 mark each for any three)
      decrease in mass of copper per minute
      increase in volume of gas per minute
      increase in depth of colour of solution / increase in copper compound per minute
      decrease in concentration of acid per minute
      decrease in pH per minute

3. Electrical conductivity (in this reaction) due to ions [1] Ions on the left but none on right [1]

Unit 8.2

LL At the start of the reaction, the volume of hydrogen given off per second is high, but as the reaction proceeds, the volume of hydrogen given off per second decreases, which is shown by the gradient of the graph decreasing.
(2 marks if all correct; 1 mark if one pair of phrases in the incorrect place)

1. a. 64 s [1]
   b. 46 cm³
   c. 33 cm³ [1]
   d. Points all plotted correctly [1] Best-fit curve through the points [1]

Unit 8.3

LL A catalyst is a substance that **increases** the **rate** of a reaction. The catalyst is **unchanged** at the **end** of the reaction. A catalyst works by providing a mechanism for the **reaction** that needs **less** energy.
(1 mark for each word in the correct place)

1. a. 24 cm³ [1]
   b. 48 cm³ [1]
   c. The smaller ones [1] More ions / particles exposed for reaction [1]
2. a. Three curves similar to those in graph in 8.2 above and levelling off [1]
      Steepest curve labelled S and shallowest curve labelled L [1]
   b.

      Horizontal axis labelled time with units of s or min [1]
      Vertical axis labelled mass with units of g [1]
      2 curves levelling off [1]
      Steeper curve labelled S / shallower curve labelled L [1]

Unit 8.4

LL In order to react, particles must **collide** with each other. The collisions must have enough **energy** to break **bonds** to allow a reaction to happen. Increasing the concentration of a reactant **increases** the **frequency** of collisions and so increases the **rate** of reaction.
(1 mark for each word in the correct place)

1. a.

      More acid particles drawn [1]
      Same number Mg particles [1]
      Particles randomly spread out including water particles [1]

Answers

b. Fewer acid particles drawn [1]
Fewer magnesium particles drawn [1]
Particles drawn randomly including water particles (and $MgCl_2$ particles) [1]

2. (When pressure is increased) molecules are closer [1] So they collide more frequently [1]

Unit 8.5

LL A with 3; B with 4; C with 1; D with 2 (2 marks if all correct, 1 mark if 2 or 3 correct)

1. a.

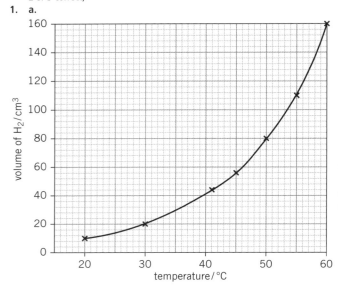

Correct axes labelled with units [1] Full use of grid [1] 7 points plotted correctly [2] (but 5 or 6 plotted correctly = [1]) Curved line of best fit [1]

b. Increases with increase in temperature [1] Comment that increase gets increasingly greater / accelerates with each 10 °C rise [1]

Unit 8.6

LL (1 mark for each word correct)

1. a. $6CO_2$ [1] $6H_2O$ [1]
b. i. As light intensity increases, rate increases [1]
ii. Some other factor is limiting / carbon dioxide is limiting / water is limiting [1]
2. $2Ag^+ + 2e^- \rightarrow 2Ag$ [1]
$2Br^- \rightarrow Br_2 + 2e^-$ [1]
Top equation is reduction **and** bottom equation is oxidation [1]

Unit 9.1

LL Reversible; Hydrated; Anhydrous; System; Dynamic; Position (1 mark each)
1. a. Reversible reaction / equilibrium reaction [1]
b. Warm / heat gently [1]

c. Heating hydrated cobalt chloride to get anhydrous cobalt chloride is endothermic [1]
So the reverse reaction must be exothermic [1]
2. Molecules randomly arranged in the mixture [1]
All three types of molecule present [1]
More molecules of hydrogen iodide than molecules of hydrogen and iodine [1]

Unit 9.2

LL If a reaction is exothermic in the forward direction, it will be **endothermic** in the reverse direction. For an exothermic reaction, when **temperature** increases, the equilibrium **shifts** in the direction of the **reverse** reaction. It **favours** the endothermic change where **heat** is taken in.
(1 mark for each word in the correct place)
1. a. left [1]
b. left [1] more [1] left [1] OR left [1] fewer [1] right [1]
NOTE: third mark dependent on second being correct
2. a. to the right [1]
b. to the left [1]
c. to the right [1]
d. no effect [1]
3. a. White precipitate disappears [1]
b. Increase in amount of white precipitate [1]

Unit 9.3

LL (1 mark for each word correct)

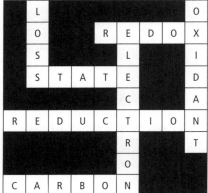

1. a. Arrow from hydrogen to water labelled oxidation [1]
Arrow from oxygen to water labelled reduction [1]
b. Arrow from lead oxide to lead labelled reduction [1]
Arrow from hydrogen to water labelled oxidation [1]
c. Arrow from iron oxide to iron labelled reduction [1]
Arrow from carbon to carbon monoxide labelled oxidation [1]
d. Arrow from carbon to carbon monoxide labelled oxidation [1]
Arrow from water to hydrogen labelled reduction [1]
e. Arrow from zinc oxide to zinc labelled reduction [1]
Arrow from carbon to carbon monoxide labelled oxidation [1]
f. Arrow from iron to iron oxide labelled oxidation [1]
Arrow from water to hydrogen labelled reduction [1]

Unit 9.4

LL A with 3, B with 4, C with 1, D with 2 (2 marks if all correct, 1 mark if 2 or 3 correct)
1. a. Oxidising agent is O_2, reducing agent is Mg [1]
b. Oxidising agent is PbO, reducing agent is H_2 [1]
c. Oxidising agent is Cl_2, reducing agent is I^- [1]
d. Oxidising agent is H_2O_2, reducing agent is I^- [1]
2. a. $Ca \rightarrow Ca^{2+} + 2e^-$ [1] oxidation [1]
b. $Cl_2 + 2e^- \rightarrow 2Cl^-$ [1] reduction [1]
c. $Al^{3+} + 3e^- \rightarrow Al$ [1] reduction [1]
d. $Fe^{2+} \rightarrow Fe^{3+} + e^-$ [1] oxidation [1]
e. $O_2 + 4e^- \rightarrow 2O^{2-}$ [1] reduction [1]
f. $Pb^{4+} + 2e^- \rightarrow Pb^{2+}$ [1] reduction [1]
g. $2Br^- \rightarrow Br_2 + 2e^-$ [1] oxidation [1]
3. From purple [1] to colourless [1]

Answers

Unit 10.1

LL (1 mark for each word)

A crossword puzzle with the following across/down words filled in:
SULFUR, O, ALKALINE, V, C, P, E, E, I, U, R, D, LITMUS, S, O, R, INDICATOR, E, L

1. A with 3, B with 5, C with 4, D with 1, E with 2
   (3 marks if all correct, 2 marks if 3 or 4 correct, 1 mark is 1 or 2 correct)
2. **a.** 1 mark each
   i. NH_3 **ii.** CH_3COOH **iii.** H_2CO_3 **iv.** H_2SO_4
   v. $Ca(OH)_2$ **vi.** H_3PO_4 **vii.** HNO_3 **viii.** $NaOH$
   b. OH [1]
3. Can 'eat away' at the surface of another substance. [1]

Unit 10.2

LL Acids react with many metals to form a salt and **hydrogen**. A salt is a substance formed when particular hydrogen **atoms** in an acid are **replaced** by a metal. Acids also react with some metal **hydroxides** to produce a **salt** and water, and with carbonates to produce a salt, **water**, and carbon **dioxide**.
(1 mark for each word in the correct place)

1. **a.** pH 13 or 9 [1] **b.** pH 3 [1] **c.** pH 7 [1] **d.** pH 3 [1]
   e. pH 9 or 13 [1]
2. **a.** zinc oxide + hydrochloric acid → zinc chloride + water [1]
   b. iron + sulfuric acid → iron sulfate + hydrogen [1]
   c. sulfuric acid + lead carbonate → lead sulfate + carbon dioxide + water [1]
   d. hydrochloric acid + tin oxide → tin chloride + water [1]
3. **a.** $Zn + H_2SO_4 \rightarrow ZnSO_4 + H_2$ [1]
   b. $MgO + 2HNO_3 \rightarrow Mg(NO_3)_2 + H_2O$ (1 for correct formula, 1 for balance)
   c. $CuCO_3 + 2HCl \rightarrow CuCl_2 + CO_2 + H_2O$ (1 for correct formula, 1 for balance)
   d. $Na_2CO_3 + 2HCl \rightarrow 2NaCl + CO_2 + H_2O$ (1 for correct formula, 1 for balance)
   e. $Ca + 2HCl \rightarrow CaCl_2 + H_2$ (1 for correct formula, 1 for balance)
   f. $2KOH + H_2SO_4 \rightarrow K_2SO_4 + 2H_2O$ (1 for correct formula, 1 for balance)

Unit 10.3

LL A base is a substance which reacts with an **acid** to form a salt. Bases that are **soluble** in water are called **alkalis**. Many metal **oxides** are basic because they react with acids to form a **salt** and water. Ammonia reacts with **sulfuric** acid to form ammonium **sulfate**; so ammonia is also a base.
(1 mark for each word in the correct place)

1. **a.** sodium hydroxide + nitric acid → sodium nitrate + water [1]
   b. calcium oxide + hydrochloric acid → calcium chloride + water [1]
   c. barium hydroxide + nitric acid → barium nitrate + water [1]
   d. hydrochloric acid + barium oxide → barium chloride + water [1]
2. **a.** $Zn + H_2SO_4 \rightarrow ZnSO_4 + H_2$ [1]
   b. $MgO + 2HNO_3 \rightarrow Mg(NO_3)_2 + H_2O$ (1 for correct formula, 1 for balance)
   c. $2NaOH + H_2SO_4 \rightarrow Na_2SO_4 + 2H_2O$ (1 for correct formula, 1 for balance)
   d. $Ca(OH)_2 + 2HNO_3 \rightarrow Ca(NO_3)_2 + 2H_2O$ (1 for correct formula, 1 for balance)
   e. $2NH_3 + H_2SO_4 \rightarrow (NH_4)_2SO_4$ (1 for correct formula, 1 for balance)
3. $Ca(OH)_2 + 2NH_4Cl \rightarrow CaCl_2 + 2H_2O + 2NH_3$ (1 for correct formula, 1 for balance)

4. Soil may be acidic (because of addition of fertilisers) [1] Calcium oxide / carbonate neutralise the acidity [1]
5. OH^- ions are formed [1]

Unit 10.4

LL Aqueous solutions of acids contain **hydrogen** ions. In strong acids **all** the acid **molecules** are dissociated (**ionised**) to form hydrogen **ions** and anions. When weak acids dissolve in **water** they become **partially** dissociated.
(1 mark for each word in the correct place)

1. **a.** pH: ethanoic = pH 2.9 [1]   hydrochloric = pH 1.0 [1]
   sulfuric = 0.7 [1]
   b. Rate of reaction: ethanoic slow [1]   hydrochloric fast [1]
   methanoic slow [1]   sulfuric fast [1]
2. **a.** $H^+(aq) + \cancel{NO_3^-(aq)} + \cancel{Na^+(aq)} + OH^-(aq) \rightarrow \cancel{NO_3^-(aq)} + \cancel{Na^+(aq)} + H_2O(l)$ [1]
   b. $H^+(aq) + OH^-(aq) \rightarrow H_2O(l)$ [1]
   c. hydrogen ions have reacted completely with hydroxide ions to form water [1]

Unit 10.5

LL A with 2, B with 4, C with 1, D with 3 (2 marks if all correct, 1 mark if 2 or 3 correct)

1. **a.** $MgO + 2HCl \rightarrow MgCl_2 + H_2O$ (1 for correct formula, 1 for balance)
   b. $SO_3 + 2NaOH \rightarrow Na_2SO_4 + H_2O$ (1 for correct formula, 1 for balance)
   c. $CuO + H_2SO_4 \rightarrow CuSO_4 + H_2O$ [1]
   d. $CO_2 + 2NaOH \rightarrow Na_2CO_3 + H_2O$ (1 for correct formula, 1 for balance)
   e. $ZnO + 2HNO_3 \rightarrow Zn(NO_3)_2 + H_2O$ (1 for correct formula, 1 for balance)
   f. $CaO + H_2SO_4 \rightarrow CaSO_4 + H_2O$ [1]
2. **a.** $SO_2 + H_2O \rightarrow H_2SO_3$ [1]
   b. $CO_2 + H_2O \rightarrow H_2CO_3$ [1]
   c. $CaO + H_2O \rightarrow Ca(OH)_2$ [1]
   d. $P_4O_6 + 6H_2O \rightarrow 4H_3PO_3$ [1]
   e. $Na_2O + H_2O \rightarrow 2NaOH$ [1]
3. $ZnO + 2KOH \rightarrow K_2ZnO_2 + H_2O$ (1 for K_2ZnO_2, 1 for H_2O)

Unit 11.1

LL Filtrate; Insoluble; Filter; Crystal; Evaporate; Heat; Oxide (1 mark each)

1. **a.** Filter off the excess zinc [1]
   b. Filter off the crystals [1]
   Wash them in the filter paper with a minimum amount of water / alcohol [1]
   Dry the crystals between sheets of filter paper / dry in a **drying** oven / dry in the air [1]
2. DBEAFC [2]
   (1 mark if one pair reversed)
3. **a.** $CaO + H_2SO_4 \rightarrow CaSO_4 + H_2O$ (1 mark for H_2SO_4, 1 mark for rest correct)
   b. $Zn + 2HCl \rightarrow ZnCl_2 + H_2$ (1 mark for HCl, 1 mark for rest correct)

Unit 11.2

LL (1 mark for each correct word)

A crossword puzzle with the following words filled in:
TITRE, I, SOLUTE, B, P, R, U, INDICATOR, P, T, E, E, ACID, T, O, T, T, NINE, END

167

Answers

1. Record the volume of acid added when the indicator just changed colour [1]
   Repeat the titration without the indicator, using the value of acid recorded [1]
   Evaporate to the point of crystallisation and leave to form crystals [1]
2. a. $NaOH + HNO_3 \rightarrow NaNO_3 + H_2O$ (1 mark for HNO_3, 1 mark for rest correct)
   b. $2NH_3 + H_2SO_4 \rightarrow (NH_4)_2SO_4$ (1 mark for H_2SO_4, 1 mark for rest correct)

Unit 11.3

LL Salts such as __nitrates__, sodium salts, and __ammonium__ salts are soluble in water. Many __carbonates / hydroxides__ and __hydroxides / carbonates__ are insoluble, except those from Group I. An insoluble substance formed when two __solutions__ of soluble __compounds__ are mixed is called a __precipitate__.
(1 mark for each word in the correct place)
1. a. soluble [1]   b. soluble [1]   c. soluble [1]
   d. insoluble [1]   e. insoluble [1]   f. insoluble [1]
2. BEADC [2] (1 mark if one pair in the incorrect order)
3. a. i. $Ag^+(aq) + \cancel{NO_3^-(aq)} + \cancel{K^+(aq)} + Br^-(aq) \rightarrow AgBr(s) + \cancel{NO_3^-(aq)} + \cancel{K^+(aq)}$ [1]
      ii. $Ag^+(aq) + Br^-(aq) \rightarrow AgBr(s)$ [1]
   b. $Ba^{2+}(aq) + SO_4^{2-}(aq) \rightarrow BaSO_4(s)$
      (1 mark for correct ions, 1 mark for state symbols)

Unit 11.4

LL When a gas is denser than __air__, you collect it by __upward__ displacement of air. If a gas is __less__ dense than air, you __collect__ it by __downward__ __displacement__ of air.
(1 mark for each word in the correct place)
1. a. C / D [1]   b. B [1]   c. A [1]
2. a. A [1]   b. D [1]
3. A with 2, B with 4, C with 1, D with 3 (2 marks if all correct, 1 mark if 2 or 3 correct)

Unit 11.5

LL A __sample__ is put on the end of a platinum __wire__ and placed at the __edge__ of a non-__luminous__ Bunsen __flame__. If __lithium__ is present, the flame is __coloured__ red. If __potassium__ is present, the flame has a lilac colour.
(1 mark for each word in the correct place)
1. $Al^{3+}(aq)$ with sodium hydroxide: white precipitate [1]
   dissolves in excess [1]
   with ammonia: white precipitate [1] insoluble in excess [1]
   $Cr^{3+}(aq)$ with sodium hydroxide: green precipitate [1]
   dissolves in excess [1]
   with ammonia: grey-green precipitate [1] insoluble in excess [1]
   $Cu^{2+}(aq)$ with sodium hydroxide: light blue precipitate [1]
   insoluble in excess [1]
   with ammonia: light blue precipitate [1]
   dissolves in excess to form a dark blue solution [1]
   $Fe^{3+}(aq)$ with sodium hydroxide: red-brown precipitate [1]
   insoluble in excess [1]
   with ammonia: red-brown precipitate [1] insoluble in excess [1]

Unit 11.6

LL A few __drops__ of nitric acid are added to the __solution__ thought to be a halide. Aqueous silver __nitrate__ is then added. If a __chloride__ is present, a white __precipitate__ is seen. If a bromide is present, a __cream__-coloured precipitate is seen.
(1 mark for each word in the correct place)
1. A with 4, B with 1, C with 3, D with 2 (2 marks if all correct, 1 mark if 2 or 3 correct)
2. a. i. $AgNO_3(aq) + NaCl(aq) \rightarrow AgCl(s) + NaNO_3(aq)$
         (1 mark for correct formulae, 1 mark for state symbols)
      ii. $Ag^+(aq) + Cl^-(aq) \rightarrow AgCl(s)$ [1]
   b. i. $BaCl_2(aq) + Na_2SO_4(aq) \rightarrow BaSO_4(s) + 2NaCl(aq)$
         (1 mark for correct formulae, 1 mark for balance, 1 mark for state symbols)
      ii. $Ba^{2+}(aq) + SO_4^{2-}(aq) \rightarrow BaSO_4(s)$ [1]

Unit 12.1

LL (1 mark for each correct word)

									M
		A			S				A
	S	r			O				G
	U				D				N
C	H	L	O	R	I	N	E		
A		F			U				S
R		U			M				I
B	O	R	O	N					U
O						A			M
N	I	C	K	E	L				

1. a. 2,8,2   2,8,3   2,8,4   2,8,5   2,8,6   2,8,7 [1]
      Al_2O_3 [1] SiO_2 [1] P_2O_3 [1]
   b. i. They increase to a maximum at Si then decrease [2]
         (They increase then decease = 1 mark)
      ii. metallic [1]
      iii. It is a giant (covalent) structure / has a lattice [1]
           All the bonds are strong / a lot of energy is needed to break all the bonds [1]
      iv. They are simple molecules [1]
          Weak forces between (the molecules) / it doesn't take much energy to overcome the weak intermolecular forces [1]

Unit 12.2

LL The __alkali__ metals have low __melting__ points and __densities__ compared with most other metals. They react with water to form a metal __hydroxide__ and __hydrogen__.
(1 mark for each word in the correct place)
1. a. Sodium: (1 mark each for any three)
      Moves rapidly over the surface / Fizzes rapidly / Melts and goes into a ball / Does not burst into flame
      Rubidium: (1 mark each for any three)
      Whizzes over the surface or moves faster than potassium / Fizzes extremely rapidly or fizzes more than potassium / Bursts into flame immediately / May explode [1]
      Melting point of potassium ALLOW between 50 and 75 °C (actual = 63 °C) [1]
      Metallic radius of rubidium = ALLOW between 0.24 and 0.27 nm (actual = 0.250) [1]
   b. Any value between 1.7 and 2.0 g/cm³ (actual = 1.88) [1]

Unit 12.3

LL When aqueous __chlorine__ is added to a __colourless__ solution of potassium bromide, the solution turns __orange__ because __bromine__ has been displaced. This is because a __more__ reactive __halogen__ displaces a __less__ reactive halogen from an aqueous solution of its __halide__.
(1 mark for each correct word)
1. a. Melting point increases down the Group [1]
   b. fluorine: gas [1]   chlorine: liquid [1]   bromine: solid [1]
      iodine: solid [1]
   c. Arrow going downwards from light to dark [1]
   d. Arrow going downwards from smaller to larger [1]
2. Orange solution turns brown [1] because iodine is formed [1]
   Bromine is more reactive than iodine [1]
   A more reactive halogen displaces a less reactive halogen from its halide [1]

Unit 12.4

LL The Group VIII gases (**noble** gases) are unreactive because their **electron** arrangements make them **stable**. It is difficult for their **atoms** to form ionic bonds by **gaining** or losing electrons, or covalent bonds by **sharing** electrons. Helium is particularly unreactive because there cannot be **more** than two electrons in the first **shell**. The other gases have **eight** electrons in their outer shell, which is a stable electronic structure.
(1 mark for each correct word)

1. A with 4, B with 3, C with 1, D with 2 (2 marks if all correct, 1 mark if 2 or 3 correct)
2. He and Ne [1]
3. a. Group VII elements have a stable electron configuration [1]
   So they cannot share electrons to form diatomic molecules [1]
   b. $Ar \rightarrow Ar^+ + e^-$ [1]

Unit 12.5

LL (1 mark for each correct word)

	I			C	O	B		
	R		S		H			
C	O	L	O	U	R	E	D	
O	N		F		O		E	
P			T		M		N		P
P			E		E		S		O
E	T	I	C			N	I		I
R						I		N
		C	A	T	A	L	Y	S	T

1. a. A, D, E, G, H (3 marks if all correct, 2 marks if 4 correct, 1 mark if 3 correct)
   b. very hard / tough / very strong [1]   form complex ions [1]
2. a. Ag^+ [1] b. Cu^{2+} c. Cr^{3+} [1] d. Fe^{3+} [1]
3. $Fe^{3+}(aq) + 3OH^-(aq) \rightarrow Fe(OH)_3(s)$
   (1 mark for correct formulae, 1 mark for balance, 1 mark for state symbols)

Unit 13.1

LL Alloys are **mixtures** of metals or mixtures of metals with non-metals. Alloys are often **harder** and stronger than pure metals. When a metal is **alloyed** with another metal, the **difference** in the sizes of the metal atoms makes the **arrangement** of the layers in the lattice less **regular**. This **prevents** the layers from sliding over each other so easily when a **force** is applied.
(1 mark for each correct word)

1. Aluminium alloy: (1 mark each for any two) low density or lightweight / doesn't corrode / strong (for weight)
   Brass: hard [1] does not corrode [1]
   Bronze: statues / ship's propellers / bearings [1] hard [1]
   Cobalt alloy: (1 mark each for any two) gas turbine blades / jet engines / high-speed drills / moving parts that rub against each other
2. Idea of not a pure metal [1] Impurities lower the melting point [1]

Unit 13.2

LL The products formed by metals that react with **cold** water are a metal hydroxide and **hydrogen**. The hydroxides are **alkaline** and so turn red litmus **blue**. The products formed by metals that only react with steam are a metal **oxide** and hydrogen. Copper does not react with water because it is **lower** in the reactivity series than hydrogen and cannot take the **oxygen** away from the hydrogen in the water.
(1 mark for each correct word)

1. a. i. $2Na(s) + 2H_2O(l) \rightarrow 2NaOH(aq) + H_2(g)$
         (1 mark for H_2, 1 mark for balance, 1 mark for state symbols)
      ii. $3Fe(s) + 4H_2O(g) \rightarrow Fe_3O_4(s) + 4H_2(g)$
         (1 mark for formulae, 1 mark for balance, 1 mark for state symbols)

2. a. Calcium, magnesium, zinc [1]
   b. Barium: gives off bubbles very rapidly with <u>cold</u> water [1]
      Barium disappears very quickly / immediately [1]
      Lead reacts slowly (when white hot) with steam / no reaction even when heated [1]
3. lead, iron, magnesium, lithium. [1]

Unit 13.3

LL Calcium is **higher** in the reactivity series than copper. So calcium atoms are better at **losing** electrons than copper. When calcium is added to **aqueous** copper(II) sulfate, calcium **displaces** the copper, and copper metal is formed. Calcium **atoms** are converted to calcium ions, which go into **solution**.
(1 mark for each correct word)

1. a. B At start: solution blue [1]
      B After 20 min: metal brown / pink [1] solution colourless / lighter blue [1]
      C After 20 min: metal silvery-grey [1]
      D At start: metal grey [1] solution blue [1]
      D After 20 min: metal brown / pink [1] solution colourless / lighter blue [1]
   b. silver < copper < iron < zinc [1]
   c. No reaction because copper is lower in the reactivity series than zinc / no reaction because zinc is higher than copper in the reactivity series [1]

Unit 13.4

LL Although aluminium is high in the **reactivity** series, samples of the metal exposed to the air do not appear to react with water or dilute **acids**. This is because **freshly** made aluminium reacts with **oxygen** in the air to form a thin **layer** of aluminium **oxide** on its surface, which is relatively **unreactive**. The oxide layer **sticks** to the metal surface strongly so is not easily removed.
(1 mark for each correct word)

1. a. reducing agent Fe and oxidising agent CuO [1]
   b. reducing agent Mg and oxidising agent Fe_2O_3 [1]
2. manganese oxide + aluminium $\rightarrow$ manganese + aluminium oxide [1]
3. a. $SnO_2 + 2C \rightarrow Sn + 2CO$
      (1 mark for correct formulae, 1 mark for balance)
   b. $2NiO + CO + H_2 \rightarrow 2Ni + CO_2 + H_2O$
      (1 mark for correct formulae, 1 mark for balance)
   c. $2PbO + C \rightarrow 2Pb + CO_2$
      (1 mark for correct formulae, 1 mark for balance)
4. a. silver < copper < tin < chromium < manganese
      (2 marks if all correct, 1 mark if 1 pair reversed)
   b. $Cr_2O_3 + 3C \rightarrow 2Cr + 3CO$
      (1 mark for correct formulae, 1 mark for balance)

Unit 13.5

LL (1 mark for each correct word)

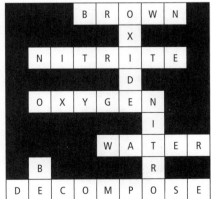

1. a. Strontium carbonate < calcium carbonate < magnesium carbonate < copper carbonate [1]
   b. The more reactive the metal, the more difficult it is to decompose the carbonate [1]
2. A with 2, B with 5, C with 1, D with 3, E with 4
   (3 marks if all correct, 2 marks if 3 or 4 correct, 1 mark if 1 or 2 correct)

Unit 14.1

LL Zinc is extracted from the ore zinc **blende**, which contains zinc **sulfide**
In one method, the ore is **roasted** in air to form zinc oxide. The zinc oxide
is then heated in a **blast** furnace with **carbon**. The carbon burns to form
carbon **monoxide**, which **reduces** the zinc oxide to zinc. Most zinc is now
produced by the **electrolysis** of zinc sulfate. This produces **purer** zinc.
(1 mark for each correct word)

1. a. Upward arrow from gold to lithium [1]
   b. Li to Al (Zn) extracted by electrolysis [1] Zn to Cu extracted by
      heating with carbon [1]
   c. i. upward arrow in 4th column [1] upward arrow in 5th column [1]
      ii. Any reasonable answer e.g. Some ores more complex than
          others / sulfur may have to be removed / other substances may
          have to be removed / some ores may require more purification
          than others [1]
   d. silver and gold / copper, silver, and gold [1]
2. $2ZnS + 3O_2 \rightarrow 2ZnO + 2SO_2$
   (1 mark for 2ZnS, 2ZnO, and 2SO$_2$, 1 mark for 3O$_2$)

Unit 14.2

LL Hematite; Reduction; Slag; Oxide; Iron; Blast; Carbon; Coke (1 mark each)

1.

B [1]

E   E [1]

A →   ← A [1]

D [1]   C [1]

2. Burning coke in air to form carbon dioxide [1]
   Reaction of carbon dioxide with coke to form carbon monoxide [1]
3. Limestone undergoes **thermal** decomposition to form calcium **oxide**.
   This reacts with **silicon** dioxide (sand), which is an **impurity** in the ore.
   The calcium **silicate** (**slag**) formed runs down the furnace and **floats**
   on top of the **molten** iron.
   (1 mark for each correct word)
4. $Fe_2O_3 + 3CO \rightarrow 2Fe + 3CO_2$
   (1 marks for correct formulae, 1 mark for balance)

Unit 14.3

LL (1 mark for each correct word)

	S	I	L	I	C	O	N
	U					X	
	L					I	
						D	
C	O	N	V	E	R	T	E	R
A						S	
R					C	A	
B	A	S	I	C		L	
O			A			B	
N	E	G	Y	X	O		

1. A with 5, B with 6, C with 1, D with 2, E with 4, F with 3.
   (3 marks for all correct, 4 or 5 correct = 2, 2 or 3 correct = 1)
2. a. $4P + 5O_2 \rightarrow 2P_2O_5$ [1]
   b. $CaO + SiO_2 \rightarrow CaSiO_3$ [1]
3. (1 mark each for any two)
   makes the steel harder / makes the steel stronger / makes the steel more
   resistant to corrosion

Unit 14.4

LL (1 mark for each correct word)

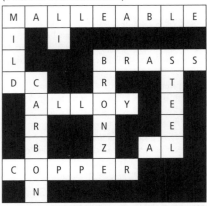

M	A	L	L	E	A	B	L	E
I		I						
L				B	R	A	S	S
D	C			R			T	
	A	L	L	O	Y		E	
	R			N			E	
	B			Z		A	L	
C	O	P	P	E	R			
	N							

1. Aluminium: (1 mark each for any two) non-toxic / resistant to corrosion /
   low density or lightweight
   Copper: Electrical wiring: good conductor of electricity [1]
   Saucepan: Good conductor of heat [1]
   Mild steel: Any two suitable uses (1 mark each) e.g. car bodies / bridges
   Property: hard / strong [1]
   Stainless steel: Use (1 mark for any two) e.g. chemical plant / cutlery /
   surgical instruments
   Property: resistant to corrosion / very hard [1]
   Tungsten steel: (1 mark for any two) e.g. resistant to wear / high melting
   point / hard

Unit 15.1

LL In a water treatment **plant**, large objects such as plant **branches** are
first trapped by metal screens. Other solid particles are then left to
settle to the bottom of the tank. The water is then passed through a
filter made of sand or gravel. This removes small **insoluble** particles.
Chlorine is added to the filtered water to kill **bacteria**, which may be
harmful to health.
(1 mark for each correct word)

1. a. Ca^{2+} [1]
   b. NO_3^- [1], SiO_3^{2-} [1], and K^+ [1] present in much higher
      concentration in river water
   c. NO_3^- [1] from fertilisers leaching into the water [1]
   d. 2.4 mg/dm^3 [1]
   e. NO_3^- nitrate [1] SiO_3^{2-} silicate [1]

Unit 15.2

LL Oxygen is separated from nitrogen by **fractional** distillation of liquid
air. When liquid air is **warmed**, nitrogen boils off **first** because it has
a **lower** boiling point than oxygen. This leaves **liquid** oxygen, which is
separated from argon, **krypton**, and xenon by further **distillation**.
(1 mark for each correct word)

1. a. Carbon dioxide and methane [1]
   b. Argon [1]
   c. Water vapour [1]
2. a. 17.1 cm^3 [1]
   b. (17.1/80) × 100 [1] = 21.4%

Unit 15.3

LL (1 mark for each correct word)

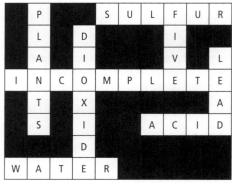

	P			S	U	L	F	U	R
	L		D				I		
	A		I				V		L
I	N	C	O	M	P	L	E	T	E
	T		X						A
	S		I			A	C	I	D
			D						
W	A	T	E	R					

Answers

1. Carbon monoxide is formed when **carbon** compounds **burn** in a **limited** supply of air. Sulfur dioxide is formed when **fossil** fuels containing **sulfur** burn in air.
   (1 mark for each correct word)
2. a. Chemical erosion / pits the limestone [1]
   b. Poisonous / toxic / stops respiration [1]
3. a. Sulfur trioxide reacts with rain water / reacts with water in the atmosphere [1] (not dissolves in rainwater)
   b. Pits the stone / chemical erosion of the stone [1]
      Limestone is calcium carbonate [1]
      Acids react with carbonates to form salts (which could be soluble), carbon dioxide (and water) [1]
   c. Acid burn on leaves / kills roots (especially of conifers) [1]
4. Lead (compounds) not added to petrol any more / modern paints don't have lead in them. ALLOW: less lead in atmosphere. [1]

Unit 15.4
LL Catalyst; Platinum; Nitric; Monoxide; Redox; Reduce; Gases (1 mark each)
1. During the morning there are more vehicles on the road [1]
   Nitrogen dioxide from vehicle exhausts goes into the atmosphere [1]
   Nitrogen dioxide concentration builds up during the day since there is still traffic on the roads [1]
   After about 12 midnight very few vehicles about so the nitrogen oxides have time to disperse into the upper atmosphere [1]
2. To turn harmful carbon monoxide [1] and nitrogen oxides [1] into nitrogen and carbon dioxide, which are not harmful (to health) [1]
3. a. (1 mark each for any two sources) Lightning / high-temperature furnaces / car exhausts / bacterial action in soil
   b. Breathing difficulties / irritates the throat / irritates the eyes [1]
4. a. $2NO_2 \rightarrow N_2 + 2O_2$ (1 mark for correct products, 1 mark for balance)
   b. $2NO + 2CO \rightarrow N_2 + 2CO_2$ (1 mark for correct products, 1 mark for balance)

Unit 15.5
LL Methane is a greenhouse gas formed by the **bacterial** decomposition of **vegetation** and as a **waste** product in the **digestive** system of animals. It is present in the **atmosphere** at a lower concentration than carbon dioxide but it **absorbs** much more heat energy per mole. It traps the **heat** in the atmosphere, which leads to **global** warming.
   (1 mark for each correct word)
1. a. Gas that absorbs and re-radiates heat energy / infrared radiation [1] in the atmosphere [1]
   b. The general trend is the same of increasing concentration of CO_2 and increasing temperature of the atmosphere [1] Reference especially to the years 1950 to 2000 [1]
   c. Between about 1895 and 1915 carbon dioxide was increasing but there was a decrease in mean temperature [1] similarly between 1940 and 1950 [1]
2. (1 mark each for any 3) e.g. rise in sea level / desertification / more extreme weather / melting glaciers / warming of sea causing death of corals etc.

Unit 15.6
LL (1 mark for each correct word)

1. a. respiration [1] release of carbon dioxide from oceans [1]
   b. photosynthesis [1] dissolving of carbon dioxide in the oceans [1]
   c. Respiration is (slightly more than) balanced by photosynthesis [1]
      The dissolving of carbon dioxide in the oceans is balanced by the release of carbon dioxide (into the atmosphere) [1]
   e. Cutting down trees / removing vegetation [1] Burning more fossil fuels [1]

Unit 15.7
LL Blocks of zinc can be placed on the hull of a ship to stop it **rusting**. Zinc is **more** reactive than **iron** so it loses **electrons** and forms **ions** more easily than iron. The zinc ions go into **solution** and so the zinc **corrodes** instead of the iron. This is called **sacrificial** protection.
   (1 mark for each correct word)
1. a. Corrosion is rapid greatly at low pH [1]
      Corrosion slows at very alkaline pH [1]
      Not much difference in corrosion between pH 3 and 9 [1]
   b. $Fe^{2+}(aq) + 2OH^-(aq) \rightarrow Fe(OH)_2(s)$
      (1 for correct formulae, 1 for balance, 1 for correct state symbols)
   c. i. Oxygen [1]
      ii. The rate of corrosion is lower at more alkaline pH values [1]
2. a. There is very little water / water evaporates very quickly [1]
   b. Idea of a layer protecting the surface of the iron [1]
      So prevents oxygen and water from reaching the iron [1]

Unit 16.1
LL For healthy growth, crop plants need three major elements: nitrogen, **phosphorus**, and potassium. Plants take up these elements in the form of nitrates, **phosphates**, and potassium **salts**. The **nitrates** are needed to make **proteins** for growth. Farmers add **fertilisers** to the soil to add back the **nutrients** that plants have absorbed for growth.
   (1 mark for each correct word)
2. a. NH_3 ammonia [1] HNO_3 nitric acid [1] NH_4NO_3 ammonium nitrate [1] H_2SO_4 sulfuric acid [1] H_3PO_4 phosphoric acid [1]
      KCl potassium chloride [1]
   b. ammonia + nitric acid $\rightarrow$ ammonium nitrate [1]
   c. i. ammonia [1] sulfuric acid [1]
      ii. potassium hydroxide / potassium carbonate [1]
         hydrochloric acid [1]
         NOT: potassium
      iii. Sodium hydroxide / sodium carbonate [1] phosphoric acid [1]
         NOT: sodium
2. a. To neutralise acid in the soil [1]
   b. $(NH_4)_2SO_4 + Ca(OH)_2 \rightarrow CaSO_4 + 2NH_3 + 2H_2O$
      (1 mark for correct formulae, 1 mark for balance)

Unit 16.2
LL Nitrogen; Iron; Catalyst; Haber; Methane; Hydrogen; Pressure (1 mark each)
1. Cracking hydrocarbons [1] Methane and steam [1]
2. $N_2(g) + 3H_2(g) \rightleftharpoons 2NH_3(g)$ (1 mark for correct formulae, 1 mark for balance)
3. a. Increasing pressure increases % yield [1]
   b. The % yield decreases with increasing temperature [1]
   c. 52% [1]
   d. Advantage: % yield higher [1]
      Disadvantage: rate of reaction lower [1]

Answers

Unit 16.3

LL (1 mark for each correct word)

		W	A	T	E	R			
			E				P	H	
	B		Z	N	S		Y		
	L				A		D		
F	E	R	T	I	L	I	S	E	R
	A			T			O		
	C		P				G		
	H		S	U	L	F	I	T	E
			L				N		
		P	A	I	N	T			

1. a. When the fuels are burnt sulfur dioxide is produced [1]
      which contributes to the formation of acid rain [1]
   b. To speed up the reaction / to lower the activation energy of the
      reaction [1]
   c. The other gases do not react with the solvent / the other gases are
      not soluble in the solvent [1]
2. a. sulfuric acid + potassium hydroxide → potassium sulfate + water [1]
      $H_2SO_4 + 2KOH \rightarrow K_2SO_4 + 2H_2O$
      (1 mark for correct formulae, 1 mark for balance)
   b. sulfuric acid + magnesium → magnesium sulfate + hydrogen [1]
      $H_2SO_4 + Mg \rightarrow MgSO_4 + H_2$ [1]

Unit 16.4

LL Vanadium; Catalyst; Trioxide; Oleum; Exothermic; Oxygen; Contact
(1 mark each)

1. a. Catalyst / to speed up the rate of reaction [1]
   b. i. As temperature increases from 300 to about 450 °C there is not
         much difference in yield / the yield gets a little less [1]
         At temperatures higher than about 450 °C the yield decreases
         markedly [1]
      ii. 92.5% (allow 92 or 93%) [1]
      iii. For an exothermic reaction the yield decreases as temperature
         increases [1]
   c. Increasing pressure shifts the equilibrium to the right [1]
      There are fewer moles / smaller volume of gas on the right in
      the equation [1]

Unit 16.5

LL Flue gas desulfurisation is the process of removing **sulfur** dioxide from
the gases formed during the **combustion** of fossil **fuels** in power
stations. The **waste** gases are passed through moist calcium **carbonate**
or calcium oxide. These compounds **neutralise** the acidic sulfur dioxide.
Solid calcium **sulfite** is formed.
(1 mark for each correct word)

1. a. T in the hole at the top [1] F near the passages to the kiln at the
      bottom [1]
   b. Through the passages into the kiln [1] It is needed to burn
      the coal / fuel [1] to provide heat (for the decomposition) [1]
   c. Limestone would decompose in the heat [1] Granite / silicon
      dioxide / aluminium oxide does not decompose in the heat [1]
   d. i. Breakdown of a substance when heated [1]
      ii. The kiln is open to the air [1] The carbon dioxide escapes [1] So
         the equilibrium shifts to the right [1]

Unit 17.1

LL A with 3; B with 4; C with 2; D with 1 (2 marks if all correct, 1 mark if
2 or 3 correct)

1. a. A family of similar compounds with similar chemical properties [1]
      due to the same functional group [1]
   b. Propene: alkenes [1] Butanol: alcohols / alkanols [1]
      Hexane: alkanes [1]
      Propanoic acid: carboxylic acids [1]

2. Ethane: C_2H_6 [1] Ethanol: C_2H_6O [1] Ethanoic acid: $C_2H_4O_2$ [1]
   Ethene: C_2H_4 [1]

3. a. b. c. d.
   (1 mark for each structure correct)

Unit 17.2

LL (1 mark for each correct word)

			F		C				
P	R	O	P	A	N	E			
			U		R		S	I	X
			R		B		E		
					O		R		
A	L	K	A	N	E		I	S	O
			L				E		
A	L	K	E	N	E	S			
			Y		T				
			L		H				

1. Compound containing only carbon and hydrogen [1]
2. a. Pentane [1]
   b. Butane [1]
   c. Hexane [1]
3. a. Propene [1]
   b. Ethene [1]
   c. Pentene [1]
4.

   (1 mark each for any 2)

Answers

Unit 17.3
LL (1 mark for each correct word)

		C					
		O		M			
		A		E			P
V	O	L	A	T	I	L	E
I				H			T
S		W	A	T	E	R	
C	U/O			N			O
O		D	I	E	S	E	L
U						R	
S	O	L	I	D		Y	

1. **a.** C [1]
   b. B [1]
2. **a.** Downward arrow [1]
   b. **i.** Easily vaporised / liquid has a low boiling point [1]
   ii. Upward arrow [1]
   c. Downward arrow [1]
   d. Upward arrow [1]

Unit 17.4
LL There is a range of **temperatures** in the distillation column, hot at the **bottom** and cooler at the **top**. Hydrocarbons with **lower** boiling points move **further** up the column and **condense** when the temperature in the column falls just below the **boiling** point of the hydrocarbons. Hydrocarbons with **higher** boiling points condense lower down the column.
(1 mark for each correct word)

1. Heat the flask with flame of constant height [1]
   Collect first fraction in test tube over a particular temperature range [1]
   Replace test tube and continue to heat until another fraction is collected over a particular temperature range [1]
   Repeat until no more fractions can be collected [1]
2. **a.** B [1]
   b. C [1]

Unit 18.1
LL Alkanes are generally **unreactive** except for the reaction with chlorine in the presence of light (a **photochemical** reaction) and **combustion**. When **excess** alkane is mixed with chlorine in a sealed tube and exposed to **sunlight**, the **green** colour of the chlorine disappears. A chlorine atom replaces a **hydrogen** atom in the alkane. This type of reaction is called a **substitution** reaction. If excess **chlorine** is present, more than one hydrogen atom is replaced by chlorine.
(1 mark for each correct word)

1. **a.** Hydrocarbons [1]
   b. Single [1] covalent [1]
2. Boiling points increase as relative molecular mass increases [1]
3. **a.** $C_5H_{12} + 8O_2 \rightarrow 5CO_2 + 6H_2O$
   (1 mark for balancing carbon dioxide and water, 1 mark for balancing oxygen)
   b. $2C_4H_{10} + 9O_2 \rightarrow 8CO + 10H_2O$
   (1 mark for balancing butane, carbon monoxide, and water, 1 mark for balancing oxygen)
   c. $CH_4 + Cl_2 \rightarrow CH_3Cl + HCl$ [1]
   d. $C_2H_6 + 2Cl_2 \rightarrow C_2H_4Cl_2 + 2HCl$ [1]
   (1 mark for HCl as product, 1 mark for balance)
4. Photochemical [1] substitution [1]

5.
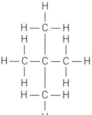

(1 mark each for any 2)

Unit 18.2
LL (1 mark for each correct word)

		C	R	A	C	K	I	N	G
		A						I	
	E	T	H	E	N	E		N	
		A						E	
		L				H			
	H	Y	D	R	O	G	E	N	
		S		X		A			
		T		I		T			
				D					
		T	H	E	R	M	A	L	

1. **a.** Residues [1]
   b. (1 mark for name and 1 mark for structure of any one)
   methane, CH_4; ethane, C_2H_6; propane, C_3H_8; butane, C_4H_{10}
   c. **i.** Gasoline/naphtha and Diesel [1]
   ii. Kerosene and fuel oil and residue [1]
3. **a.** $C_{10}H_{22} \rightarrow C_4H_{10} + C_6H_{12}$ [1]
   b. $C_{14}H_{30} \rightarrow C_3H_8 + C_4H_8 + C_7H_{14}$ [1]

Unit 18.3
LL We can tell the difference between an unsaturated and a **saturated** compound by adding **aqueous** bromine to a sample of the compound. Aqueous bromine is **orange** in colour. If the aqueous bromine is **decolourised**, the compound is **unsaturated**. If the bromine water **remains** orange, the compound is saturated.
(1 mark for each correct word)

1. **a.** Alkenes: propene [1] butene [1]
   Molecular formulae: C_2H_4 [1] C_4H_8 [1] C_5H_{10} [1]
   Boiling points: propene −60 to −40 °C (actual −48 °C) [1]
   Hexene 50 to 80 °C (actual 63 °C) [1]
   b. pentene and hexene [1]
2. **a.** C=C ringed [1]
3. A [1]   B $H_2O(g)$ (NOT $H_2O(l)$)

Answers

Unit 18.4

LL (1 mark for each correct word)

1. a. Arrow under flask [1]
   b. Oxidising agent [1]
   c. Purple [1] to colourless [1]
   d. To condense the vapours / to stop loss of vapour [1] of ethanol / ethanoic acid [1]

2. a. $C_2H_5OH + 3O_2 \rightarrow 2CO_2 + 3H_2O$
      (1 mark for correct formulae, 1 mark for balance)
   b. $C_2H_5OH + 2[O] \rightarrow CH_3COOH + H_2O$
      (2 marks if all formula correct, 1 mark if two of C_2H_5OH, CH_3COOH, H_2O correct)

Unit 18.5

LL Esters are used in **flavourings** and perfumes. Esters are made by heating an **alcohol** with a carboxylic acid. A few drops of **sulfuric** acid are added to act as a **catalyst**. After the reaction is complete, excess sodium carbonate is added to **neutralise** any excess acid. Esters are **insoluble** in water, so they can be separated from the rest of the reaction mixture using a **separating** funnel.
(1 mark for each correct word)

1. a. The reaction is an equilibrium reaction [1] Both unionised acid molecules and (ethanoate) ions are present [1]
   b. It is accepting a proton [1] from ethanoic acid [1]

2. a. $2CH_3COOH + 2Na \rightarrow 2CH_3COO^-Na^+ + H_2$
      (1 mark for formula of each product ALLOW CH_3COONa, 1 mark for balance)
   b. $2CH_3COOH + Mg \rightarrow (CH_3COO^-)_2Mg^{2+} + H_2$
      (1 mark for formula of each product ALLOW $(CH_3COO)_2Mg$, 1 mark for balance)
   c. $CH_3COOH + NaOH \rightarrow CH_3COO^-Na^+ + H_2O$
      (1 mark for each correct product)
   d. $CH_3COOH + CH_3OH \rightarrow CH_3COOCH_3 + H_2O$
      (1 mark for each correct reactant and 1 mark for H_2O)

3. a. [1]
   b. [1]

4. a. butyl methanoate [1]
   b. ethyl propanoate [1]

Unit 19.1

LL A polymer is a substance that has very large **molecules** formed when lots of small molecules called **monomers** join together. This process is called **polymerisation**. When poly(ethene) is formed, one of the C=C **bonds** of **ethene** is broken and the monomers **join** together in a chain.
(1 mark for each correct word)

1. 8 carbon atoms in a chain with two hydrogen atoms attached to each [1]
   Continuation bonds shown on the end carbon atoms [1]
2. They cannot be broken down / be decomposed [1] by living organisms / bacteria / fungi [1]
3. a. Saves raw material / saves energy / can be made into new objects / could be cracked to different chemicals [1]
   b. Plastics do not have to be sorted / cheap [1]
   c. Heat can be used to generate electricity (via turbine) / saves burning fossil fuels / heat can be used directly e.g. to heat greenhouses [1]

Unit 19.2

LL When **monomers** containing C=C double **bonds** are polymerised, no other molecule is **formed** apart from the polymer. We call this type of polymerisation **addition** polymerisation. An addition reaction is a reaction in which **two** or more molecules **combine** and **no** other molecule is formed.
(1 mark for each correct word)

1.
   (1 mark for correct number of C atoms, 1 mark for rest of structure, 1 mark for continuation bonds on end carbon atoms)

2.
   (1 mark for structure, 1 mark for brackets and n, 1 mark for continuation bonds on end carbon atoms)

3. A double bond [1] rest of structure correct [1]
   B double bond [1] rest of structure correct [1]

Unit 19.3

LL In condensation polymerisation, molecules with different **functional** groups react together. A **small** molecule such as **water** or hydrogen **chloride** is **eliminated**.
(1 mark for each correct word)

1. a.
   b.
      (1 mark each)
   b. i. ester [1] ii. amide [1]
2. a. Alcohol [1] Carboxylic acid [1]
   b. −OH and −COOH groups [1] rest of molecule correct [1]

Unit 20.1

LL (1 mark for each correct word)

1. a. Correct repeat unit e.g. first bracket between the first N and CHR from the left, second bracket between the next N and CHR to the right [1]
    b. Amide / peptide [1]

2.

$$H_2N-\overset{\displaystyle H}{\underset{\displaystyle H}{\overset{|}{\underset{|}{C}}}}-C\overset{\displaystyle O}{\underset{\displaystyle OH}{<}}$$

COOH and NH_2 groups [1] rest of structure correct [1]

3. Breakdown of a substance [1] using water (or acid or alkali as catalysts) [1]

Unit 20.2

LL The different carbohydrates in a **mixture** can be identified using **paper** chromatography. After carrying out chromatography, the position of the solvent **front** is marked. The **chromatography** paper is then sprayed with a **locating** agent. This makes the **position** of each carbohydrate visible as a **coloured** spot.
(1 mark for each correct word)

1. a. 6 [1]
    b. condensation [1]
    c. HO─☐─OH [1]

2. ─O─⬠─O─⬠─O─⬠─O─⬠

    (1 mark for 4 repeat units, 1 mark for correct structure, 1 mark for continuation bonds)

3. a. In at the bottom (←), out at the top (←) [1]
    b. To stop vapour escaping [1] to stop HCl escaping [1]

Unit 20.3

LL Anaerobic; Ethanol; Enzymes; Yeast; Glucose; Distil (1 mark each)

1. a. Fermentation: reagents glucose [1] temperature ALLOW between 10 and 40 °C [1]
        Pressure atmospheric / 1 atm [1] Catalyst enzymes / yeast [1]
        Hydration: reagents ethene and steam [1] temperature 500–600 °C [1]
        Pressure 60–70 atmospheres [1] catalyst phosphoric acid [1]
    b. (1 mark each for any two) takes a long time / ethanol is dilute / need to distil off the ethanol / batch process is inefficient / lot of waste
    c. (1 mark each for any two) Uses renewable resources / relatively cheap / does not require **very** high temperature and pressure
    d. (1 mark each for any two) Reaction is fast / reaction can be run continuously / gives pure ethanol or atom economy is (nearly) 100%

Unit 21.1

1. A with 5, B with 8, C with 1, D with 6, E with 7, F with 2, G with 3, H with 4
    (4 if all correct, 3 if 6 or 7 correct, 2 if 4 or 5 correct, 1 if 2 or 3 correct)

2. I with 11, J with 16, K with 9, L with 13, M with 10, N with 15, O with 12, P with 14
    (4 if all correct, 3 if 6 or 7 correct, 2 if 4 or 5 correct, 1 if 2 or 3 correct)

3. Q with 22, R with 23, S with 17, T with 18, U with 19, V with 20, W with 21
    (4 if all correct, 3 if 6 correct, 2 if 4 or 5 correct, 1 if 2 or 3 correct)

Unit 21.2

1. a. Hydrocarbons are **compounds** containing **only** hydrogen and **carbon** atoms. [1]
    b. only [1]

2. a. A compound is a **substance** that contains **two** or more **different** atoms **bonded** together (2 marks if all correct, 1 mark if 1 or 2 correct)
    b. different [1] bonded (or joined) [1]

3. a. Relative atomic mass is the **average** mass of naturally occurring **atoms** of an **element** on a **scale** on which the ^{12}C atom has a mass of exactly **twelve** units.
        (3 marks if all correct, 2 marks if 3 or 4 correct, 1 mark if 1 or 2 correct)
    b. Relative atomic mass is the average mass of naturally occurring atoms of an element on a scale on which the ^{12}C atom has a mass of exactly twelve units.
        (1 mark each correct word)

4. A mole is the **amount** of substance that has the same number of **particles** (atoms, ions, **molecules**, or electrons) as there are **atoms** in exactly twelve **grams** of the carbon-12 **isotope**.
    (3 marks if all correct, 2 marks if 4 or 5 correct, 1 mark if 2 or 3 correct)

Unit 21.3

1. 'minimum' not crossed out [1]
2. type / kind / sort [1]
3. 'atoms' not crossed out [1] 'element' not crossed out [1]
4. decomposition / breakdown [1] electricity / electric current [1]
5. 'loses' not crossed out [1]
6. 'ions' not crossed out [1]
7. increases [1] 'volume' not crossed out [1] 'frequency' not crossed out [1]
8. 'ions' not crossed out [1] move [1]
9. 'atom' not crossed out [1] 'atom' not crossed out [1]
10. 'components' not crossed out [1] 'physical' not crossed out [1]
11. 'double' not crossed out [1] 'orange' not crossed out [1] 'colourless' not crossed out [1]

Unit 21.4

1. A with 5, B with 3, C with 6, D with 1, E with 4, F with 2
    (3 marks if all correct, 2 marks if 4 or 5 correct, 1 mark if 2 or 3 correct)

2. A with 7, B with 6, C with 1, D with 5, E with 2, F with 3, G with 4
    (4 marks if all correct, 3 marks if 5 or 6 correct, 2 marks if 3 or 4 correct, 1 mark if 1 or 2 correct)

3. A with 6, B with 4, C with 5, D with 1, E with 3, F with 2
    (3 marks if all correct, 2 marks if 4 or 5 correct, 1 mark if 2 or 3 correct)

Unit 21.5

1. A with 5, B with 7, C with 1, D with 3, E with 2, F with 4, G with 6
    (4 marks if all correct, 3 marks if 5 or 6 correct, 2 marks if 3 or 4 correct, 1 mark if 1 or 2 correct)

2. Diamond has a high melting point ..... because it takes a lot of energy ..... to break the strong ..... covalent bonds that exist ..... between all the carbon atoms.
    (2 marks if all correct, 1 mark if one pair incorrect)

3. The higher the temperature, ... the faster the particles move ..... because they have more energy and collide ..... with a greater frequency. The number of particles having energy ..... equal to or greater than the activation energy ..... also increases, so there is ..... more chance of collisions between ..... reactant particles being successful.
    (1 mark for first sentence correct, 2 marks for second sentence correct – but 1 mark if one pair incorrect in second sentence)

Answers

Unit 21.6

1. a. To remove carbon monoxide and nitrogen oxides [1] which are harmful to health [1]
   OR To change carbon monoxide and nitrogen oxides to carbon dioxide and nitrogen [1] which are not harmful to health [1]
   b. Chemical reaction that needs light to occur [1]
   c. CHO [1]
   d. layer of cold air below layer of hot air [1]
   e. 1 mark each for any two of irritates eyes / breathing difficulties / worsens asthma
   f. high temperature [1] high pressure [1]
   g. 3 [1]
2. Boils, explain, motion, energy, particles (1 mark each)

Unit 21.8

1. a. two [1] isotopes [1] industry [1]
   b. Describe [1] separate [1] dyes [1]
   c. Describe [1] observe [1] acid [1] zinc [1]
   d. Describe [1] explain [1] condensation [1] energy [1] movement [1] particles [1]
2. a. The word 'all' has not been considered [1] Metals such as potassium are soft and have relatively low melting points [1]
   b. No observation has been made [1] Observations are what you see (or hear or feel) / hydrogen gas is naming a substance (not an observation) [1]
   c. This is not an industrial use [1] An industrial use would be e.g. measuring thickness of paper / a use in medicine has been given [1]
   d. This is not a feature of the molecule [1] The presence of a C=C double bond is a feature of the molecule / A test has been given instead [1]

Unit 22.2

1. a. Use [1] suggest [1]
   b. Describe [1] give [1]
   c. Use [1] deduce [1]
   d. Describe [1] explain [1]
   e. Draw [1] determine [1]

Unit 22.4

1. Any three suitable (1 mark each), e.g. uses of particular chemicals / properties of selected elements / chemical tests, e.g. water, ions, unsaturation / types of chemical reaction, e.g. neutralisation, condensation, addition, photochemical
2. Any 4 suitable with result (1 mark each), e.g.
   + metal hydroxides → salt + water
   + metal oxide → salt + water
   + carbonate → salt + carbon dioxide + water
   + methyl orange → turns red / pink
   + ammonia → ammonium salt formed
   Taste → sour (although this is not recommended!)
   What makes them acid? → hydrogen ions
3. (1 mark for each reaction or property and 1 mark for each correct result up to 12) e.g.
   Alkane + chlorine in presence of light → chloroalkane + HCl
   Alkane + (excess) oxygen/air → carbon dioxide + water
   Alkane + limited oxygen → carbon monoxide + water
   Alkane (heat with Al_2O_3) → mixture of alkanes and alkenes
   Alkene + bromine water → bromine water decolourised
   Alkene + hydrogen (in presence of Ni catalyst) → alkane
   Alkene + steam (in presence of catalyst) → alcohol
   Alkene + (excess) oxygen / air → carbon dioxide + water

Unit 22.6

1. A low melting point [1]
   B / C / D high melting point [1] conduct electricity when molten [1]
   soluble in water [1]
   E or I metal [1] giant covalent structure [1]
   For the metal (E or I) (1 mark each for any 3) conduct electricity or conduct heat / ductile / malleable / lustrous (shiny) IGNORE: high melting point / sonorous
   For the giant covalent structure (E or I) (1 mark each for any 3) high melting point / generally do not conduct electricity / insoluble in water

2. A increase in rate [1] B concentration [1] C increase rate [1]
   D the particles collide with greater frequency [1]
   E Increases rate [1] F surface [1] G increase rate [1]
   H increased number of particles exposed for collisions / increased collision frequency [1]

Unit 23.1

1. a. 2 Na and 1 O [1]   b. 3 Mg and 2 N [1]
   c. 1 P and 3 Cl [1]   d. 2 Al and 3 O [1]
   e. 2 H, 1 S, and 4 O [1]
2. a. 4 H and 2 S [1]   b. 15 Mg and 10 N [1]
   c. 6 Al and 9 S [1]   d. 6 Fe and 8 O [1]
   e. 6 Li, 3 C, and 9 O [1]
3. a. $(2 × 27) + (3 × 16) = 102$ [1]
   b. $(2 × 23) + 12 + (3 × 16) = 106$ [1]
   c. $207 + 32 + (4 × 16) = 303$ [1]

Unit 23.2

1. a. 1 Sn, 2 S, 8 O [1]   b. 2 N, 8 H, 1 S, 4 O [1]
   c. 1 Ni, 2 Cl, 8 O [1]   d. 1 Ba, 2 I, 6 O [1]
2. a. $119 + (2 × 32) + (8 × 16) = 311$ [1]
   b. $(2 × 14) + (8 × 1) + 32 + (4 × 16) = 132$ [1]
   c. $59 + (2 × 35.5) + (8 × 16) = 258$ [1]
   d. $137 + (2 × 127) + (6 × 16) = 487$ [1]
3. $59 + (2 × 35.5) + 6 × (2 × 16) = 238$ [1]

Unit 23.3

1. a. actual yield = $\frac{\% \text{ yield}}{100}$ × theoretical yield [1]
   b. theoretical yield = $\frac{\text{actual yield}}{\% \text{ yield}}$ × 100 [1]
2. a. moles = concentration (in mol/dm^3) × volume (in dm^3) [1]
   b. volume (in dm^3) = $\frac{\text{moles}}{\text{concentration (in } mol/dm^3)}$ [1]
3. mass = density × volume

Unit 23.4

1. a. $1 × 10^6$ [1]   b. 70 000 [1]
   c. $3.3 × 10^3$ [1]   d. 3200 [1]
2. a. $1 × 10^{-5}$ [1]   b. 0.005 [1]
   c. $3.5 × 10^{-3}$ [1]   d. 0.25 [1]
3. a. 1.4 [1]   b. $3.6 × 10^{-5}$ [1]
4. a. $1.14 × 10^{-5}$ [1]   b. $2.67 × 10^2$ [1]

Unit 23.5

1. a. 87.7 % [1]
   b. mass pure substance = 3.45 – 0.12 kg = 3.33 kg [1]
   % purity = 96.5 % [1]
2. a. i. 6 [1]
   ii. 96 cm^2 [1]
   b. $5 × 5 × 5 = 125$ cm^3 [1]

Unit 23.6

1. a. i. 4.36 [1]   ii. 0.0873 [1]
   iii. 137 [1]   iv. 0.00550 [1]
   b. i. 440 [1]   ii. 3.4 [1]
   iii. 57 [1]   iv. 0.0055 [1]
2. mol pentane 0.0926388 [1]   rounded 0.09 [1]
   × 5   0.4631944 [1]   rounded 0.5 [1]
   × 24   11.1   [1]   13.2 [1]

Unit 23.7

1.

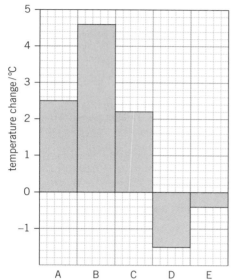

Axes labelled correctly [1] Full use of graph paper [1]
Positive values correct [1] Negative values correct [1]

Unit 23.8

1. a. Line not continued to 0-0 point [1] Line not a continuous curve [1]
        Line has more points above it than below it [1]
    b. Points not clear [1] No units on *y*- or *x*-axis [1] Full grid not used / lot of grid is space [1]
    c. Anomalous point included in the line [1] Straight lines between points / not a smooth curve [1]
2. a. 34 cm³ [1]   b. 44 cm³ [1]

Unit 23.9

1.

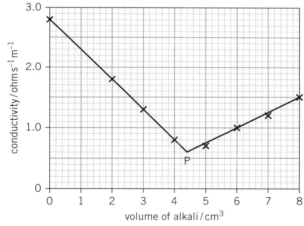

    a. Axes correctly labelled [1]
        Points all correct [2] (1 mark if one point incorrect or missing)
    b. Lines correct (two straight intersecting lines) [1]
        Lines intersect between 4 and 5 cm³ and P labelled [1]
    c. P is 4.4 cm³ [1]

2.

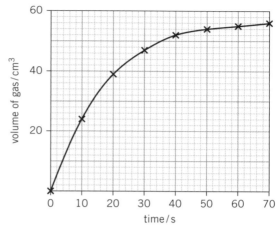

Axes correctly labelled and full grid used [1]
Points all correct [2] (1 mark if one point incorrect or missing)
Smooth curve between the points [1]

Unit 23.10

1.

    a. Axes correctly labelled and full grid used [1]
        Points all correct [2] (1 mark if one point incorrect or missing)
        Smooth curve between the points [1]
    b. 0.16/40 [1] 4 × 10⁻³ g/s [1]
    c. The line starts to curve / the rate is not constant [1]

Unit 24.1

1. Goggles / safety glasses / eye protection [1]
2. Carry out in a fume cupboard / hood [1] Use gloves / lab coat / eye protection [1]
3. A Explosive [1] B Corrosive [1] C Toxic / Poisonous [1] D Flammable [1]
    E Harmful [1]
4. a. Flammable [1]   b. Corrosive [1]   c. Toxic / Poisonous [1]
5. Funnel has no tap [1] Gas would escape from funnel [1]
    Tube of water [1]
    Sulfur dioxide is soluble in / reacts with water [1]
    Gas jar wrong way up [1] Hydrogen chloride is heavier than air [1]

Unit 24.2

1. Apparatus to be used [1] Quantities or concentrations to be used [1]
    Conditions to be used, e.g. heat [1] Measurements or observations to be made [1] What to control and what to vary (idea of fair test) [1]
2. a. Temperature [1]
    b. Volume of carbon dioxide [1]
    c. (1 mark each for any two) time / mass of calcium carbonate / surface area of calcium carbonate / concentration of acid.
3. a. Mass of fuel burnt [1]
    b. Temperature rise [1]
    c. (1 mark each for any two) distance of burner from can / same copper can / same volume of water in can / same height of flame or same wick used in burner

Unit 24.3

1. Repeating the results several times until consistent results are obtained [1]

   Using equipment that measures accurately, e.g. burette instead of measuring cylinder [1]

2.

volume	temperature
0.2	21
4.2	23
8.6	27
12.4	31

   (For each column: 2 marks if all correct, 1 mark if 2 or 3 correct)

3. S = 34 cm³ [1] T = 15 cm³ [1]

Unit 24.4

1.

time / s	0	10	20	30	40	50	60	70	80	90
mass change / g (first set)	0.0	0.5	0.89	1.24	1.44	1.56	1.68	1.79	1.82	1.82
mass change / g (second set)	0.0	0.72	1.04	1.24	1.44	1.52	1.62	1.74	1.78	1.78
average / g	0.0	0.61	0.965	1.24	1.44	1.54	1.65	1.765	1.80	1.80

   (1 mark for time, masses and average in column 1, 1 mark for correct units, 1 mark for correct values, 1 mark for correct averages)

2.

temperature / °C	time / s	$\frac{1}{\text{time}}$ / s⁻¹

   (1 mark for temperature, time and rate, 1 mark for correct units)

Unit 24.5

1. a. Independent: Time [1] Dependent: Conductivity [1]
   b. Independent: Concentration of acid [1] Dependent: Time [1]

2. a.

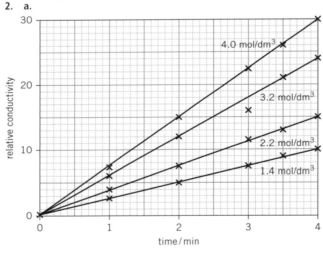

   Lines correct (1 mark each) [4] Axes of graph correctly labelled [1]
   b. The one at 3 minutes for the concentration of 3.2 mol/dm³ [1]
      It does not fall on the same line as the other points / It is lower than expected [1]

Unit 24.6

1. a. i. A, B, and E [1]
      ii. Doubling the concentration doubles the rate / rate is directly proportional to concentration [2]
         OR Increasing the concentration increases the rate (alone) [1]
   b. i. B, C, and D [1]
      ii. Doubling the concentration doubles the rate / rate is directly proportional to concentration [2]
         OR Increasing the concentration increases the rate (alone) [1]

---

   c. With only 2 different concentrations, a consistent pattern cannot be found. [1]
      Idea that it needs at least 3 points to show proportionality / show doubling concentration doubles the rate [1]

2. Deduce the gradient of each line [1] From the slope of relative conductivity divided by time [1] Plot rate against concentration (to give straight line through origin) [1]

Unit 24.7

1. a. As pressure increases, volume decreases [1]
      The rate of decrease decreases with increasing volume ALLOW: volume is inversely proportional to pressure / the graph is a concave downwards curve [1]
   b. Volume of gas is proportional to concentration of acid [2]
      (If 2 marks not scored: Volume of gas increases linearly with concentration of acid = 1 mark)
   c. Rate doubles for every 10 °C rise [2]
      (If 2 marks not scored: rate gets faster as temperature increases = 1 mark)

Unit 24.8

1. a. Thermometer [1] Beaker [1] Top-pan balance and weighing boat [1] Water bath with temperature control ALLOW: Bunsen burner and tripod and gauze [1] Stirring rod [1]
   b. Independent variable: temperature of the water [1]
      Dependent variable: mass of solid added [1]
   c. (1 mark each for any two) Volume of the water / rate at which heat is applied e.g. from water bath or Bunsen / rate of stirring

2. Place 50 cm³ / 100 cm³ water in beaker [1]
   Heat water to fixed temperature [1]
   Add small weighed amounts of solid potassium chloride to the water and stir [1]
   Keep adding small weighed amounts until no more dissolves [1]
   Repeat at different temperatures [1]

3. a. (1 mark each for any two) If using Bunsen, the water will cool while the substance is being added / if using water bath, temperature control depends on sensitivity of thermostat / heat may be given out or absorbed when substance dissolves in water.
   b. (1 mark each for any two) Use a temperature-controlled water bath OR more sensitive temperature control / use larger volume of water so that inaccuracies due to adding small amounts of solid are reduced / use insulated container so heat losses reduced / add solid to the water until saturated solution formed, then allow to cool and take temperature when crystals first appear.
   c. Temperature change [1], because thermometer reading precise to 0.5 °C at best and temperature change is only a few degrees, whereas mass can be measured to a precision of two or three decimal places [1]

Unit 24.9

1. a. Burette / measuring cylinder / gas syringe [1] Flask [1] Stopper/ bung and connecting tubing [1] Stopclock [1] Top pan-balance and weighing boat [1]
   b. Independent variable: Either: (i) Time (to collect given volume of gas) or (ii) Volume of gas (collected in a given time) [1]
      If (i) volume of gas / If (ii) time [1]
   c. Mass / moles of each metal oxide [1]
      Concentration of hydrogen peroxide [1]
      Volume of hydrogen peroxide / temperature / size of metal oxide particles [1]

2. Small amount of oxide / stated values e.g. 0.1 to 1 g [1] Suitable volume of hydrogen peroxide e.g. 1–50 cm³ [1] Add to flask and start stopclock [1] Measure volume of gas collected in fixed time / Measure time (to collect fixed volume of gas [1] Repeat under same conditions with another metal oxide [1]

3. a. Difficult to measure volume accurately when liquid level / syringe plunger moving [1]
      Difficult to read stopclock and volume at same time [1]
   b. Get someone else to read the stopclock or volume [1]
      Use more sensitive balance (reading to 3 or 4 decimal places) [1]

Answers

Unit 26.1

a. 2,8,5 [1]

b. 16 [1]

c. Does not conduct electricity / does not conduct heat [1]
   Low melting point / low boiling point [1]

d. P_4 (or 4P) + $5O_2 \rightarrow 2P_2O_5$
   (1 mark for correct formulae, 1 mark for balance)

e. PO_4^{3-} [1]

f. Nitrate / NO_3^- [1]

g. Fertilisers needed for plant growth / for plant proteins [1]
   Sources of nitrogen / phosphorus / potassium in soil used up by growing plants [1]

h.

   1 pair of electrons shared between each of the 3 H atoms and the central P atom [1]
   Lone pair on the P atom [1]

Unit 26.2

a. Double C=C bond [1]

b. Bromine water / bromine ALLOW acidified potassium manganate(VII) [1]
   Decolourised [1]

c. i. Reduction is gain of electrons [1]
   ii. High temperature [1] High pressure [1] Catalyst [1]
   iii.

   (2 marks for full structure, 1 mark if OH drawn instead of O − H)

d. i. Grind up the onion leaves in a solvent / water / alcohol [1]
      Filter [1]
   ii. Chromatography [1]

Unit 26.3

a. CsCl / Cs$^+$Cl$^-$ [1]

b. Giant structure / ionic structure [1]
   All the bonds are strong / strong electrostatic forces between all ions [1]
   (Mention of atoms / intermolecular forces = maximum 1 for question)

c. The ions are free to move (from place to place) [1]

d. i. Conduct electricity [1] unreactive / inert [1]
   ii. Anode: $2Cl^- \rightarrow Cl_2 + 2e^-$
      (1 mark for formulae, 1 mark for balance)
      Cathode: $Cs^+ + e^- \rightarrow Cs$ [1]

e. Mol Cs = 5.32/133 = 0.04 mol [1]
   Actual yield of CsCl = 6.4/168.5 = 0.038 [1]
   % yield = 0.038/0.04 = 95% [1]
   (or calculation based on masses)

Unit 26.4

a. 6.4–6.5 min [1]

b. i. 16 cm^3 [1]
   ii. 27/2 = 13.5 cm^3/min [1]

c. Initial gradient steeper [1] Ends up at the same volume of gas [1]

d. Increasing concentration increases the number of particles per unit volume / particles closer together [1]
   Frequency of collisions increases / number of collisions per second increases [1]

e. Faster because greater surface area of powder [1]
   More particles of magnesium exposed to hydrochloric acid [1]

Unit 26.5

a. Decomposition [1] endothermic [1]

b. Bubble through limewater [1] limewater turns milky / cloudy [1]

c. i.

   Axes correctly labelled [1] Points plotted correctly [1]
   Curve of best fit drawn [1]
   ii. Mass of CO_2 from graph = 3.0 g [1]
      Moles CO_2 = 3.0/44 = 0.068 mol [1]
      Volume = 0.068 × 24 = 1.64 dm^3 [1]

Unit 26.6

a. i. Arrow under the flask [1]
   ii. A (round-bottomed) flask [1] B gas jar [1]
   iii. To dry the ammonia / To remove water [1]
   iv. Put damp red litmus beneath gas jar [1]
      Full when (litmus) turns blue [1]

b. $(NH_4)_2SO_4 + 2NaOH \rightarrow 2NH_3 + Na_2SO_4 + 2H_2O$
   (1 mark for correct formulae, 1 mark for balance)

c.

   (1 mark for bonding pairs of electrons, 1 mark for the lone pairs on each nitrogen atom)

Unit 26.7

a. Circle around the O–H group (not COOH group) [1]

b. i. Carbon, hydrogen, and oxygen [2] (1 mark for any two of these)
   ii. Ethanol [1]

c. Carbon dioxide [1] Water [1]

d. Filtration [1]

e. i. Ester [1]
   ii. No continuous carbon chain [1]
      Idea of the COO groups being formed by condensation reactions [1]

f. Purple [1] to colourless [1]

Unit 26.8

a. i. Helium and neon [1]
   ii. ALLOW: values between 0.08 and 0.1 [1]
   iii. Gas [1], −118 °C is above the boiling point [1]
   iv. Increases down the Group [1]

b. i. brown ALLOW: grey / black [1]
   ii. iodide is being converted to iodine / oxidation number of iodine increases [1] oxidation number of Xe decreases [1]
   iii. mol XeF$_4$ = 8.28/207 = 0.04 mol [1]
      0.04 mol Xe [1]
      0.04 × 24 = 0.96 dm^3 Xe [1]

Answers

Unit 26.9

a.   i.   Car exhausts / High temperature furnaces / Lightning   [1]

     ii.   Acid rain / Kills trees / Acidifies lakes / Erodes limestone / Corrodes metal structures etc.   [1]

     iii.   Proximity: Far apart [1] Motion: fast / random   [1]

b.   i.   Colour gets lighter [1] Position of equilibrium moves to the left   [1]

     In direction of fewer gas molecules / fewer moles in the equation   [1]

     ii. $NO_2 = 46$ [1] $N_2O_4 = 92$ [1]

     iii.   Entirely NO_2 at 140 °C / more NO_2 at higher temperature   [1]

     The higher the temperature the more the equilibrium goes to the right   [1]

     For an endothermic reaction the position of equilibrium moves to the right with increase in temperature / increase in temperature favours the endothermic reaction   [1]

c. $2NO_2 \rightarrow 2NO + O_2$

     (1 mark for correct formulae, 1 mark for balance)

Unit 26.10

a.   Any suitable indicator, e.g. methyl orange / litmus / phenolphthalein   [1]

b.   Potassium sulfate   [1]

c.   i. $(12.5/1000) \times 0.2 = 2.5 \times 10^{-3}$ mol   [1]

     ii. 5.0×10^{-3} mol   [1]

     iii. $5.0 \times 10^{-3} \times 1000/25 = 0.20$ mol/dm^3 [1]

d. $H^+ + OH^- \rightarrow H_2O$ [1]

e.

```
      H   O        H   H   H   H
      |   ||       |   |   |   |
  H — C — C — O — C — C — C — C — H
      |            |   |   |   |
      H            H   H   H   H
```

     (2 marks if all correct, 1 mark if ester group shown as COO)